Encyclopedia of Salmonella: Advanced Concepts

Volume II

Contents

Preface

Salmonella have continued to exist for more than 100 years now. Researchers and experts have been tirelessly making efforts to control salmonellae since the discovery of salmonella in 1885. A few of the known serovars are responsible for causing majority of foodborne outbreaks. While certain advancements have been made in this field, it is yet to be eliminated from developed and developing countries. This book is a collection of contributions made by researchers and experts from all over the world. It focuses on a variety of topics related to understanding and controlling this pathogen. Some of the topics are: virulence and pathogenicity; salmonella-invasion, evasion & persistence; and molecular technology for salmonella detection.

The information contained in this book is the result of intensive hard work done by researchers in this field. All due efforts have been made to make this book serve as a complete guiding source for students and researchers. The topics in this book have been comprehensively explained to help readers understand the growing trends in the field.

I would like to thank the entire group of writers who made sincere efforts in this book and my family who supported me in my efforts of working on this book. I take this opportunity to thank all those who have been a guiding force throughout my life.

Editor

Virulence Characterization of *Salmonella* Typhimurium I,4,[5],12:i:-, the New Pandemic Strain

Madalena Vieira-Pinto[1], Patrícia Themudo[2], Lucas Dominguez[3],
José Francisco Fernandez-Garayzabal[3,4], Ana Isabel Vela[3,4],
Fernando Bernardo[5], Cristina Lobo Vilela[5] and Manuela Oliveira[5]
*[1]Departamento das Ciências Veterinárias, CECAV Laboratório de TPA &
Inspecção Sanitária, Universidade de Trás-os-Montes e Alto Douro
[2]Laboratório Nacional de Investigação Veterinária, Lisboa
[3]Centro de Vigilancia Sanitaria Veterinaria (VISAVET),
Universidad Complutense, Madrid
[4]Departamento de Sanidad Animal, Facultad de Veterinaria,
Universidad Complutense, Madrid
[5]Centro Interdisciplinar de Investigação em Sanidade Animal,
Faculdade de Medicina Veterinária da Universidade
Técnica de Lisboa, Lisboa
[1,2,5]Portugal
[3,4]España*

1. Introduction

It is really impossible to estimate the volume of ink that has been spent writing about *Salmonella* since its first description in 1885, by Daniel D. Salmon. Nowadays, performing a search on this bacterial genus using web databases originates more than 17 million references. Before the web era, the dissemination of information on *Salmonella* was extremely complex or even impossible, and this is probably the reason why *Salmonella* taxonomy has been so difficult to establish during the first century after its description. The taxonomy of this bacterial genus still today remains under revision.

Sadly, *Salmonella* is the paradigm of a popular microbe by quite regrettable reason: from the most common citizen to the most qualified microbiologist, everyone has already talked about *Salmonella*. The reason for this is, surely, its incrimination on severe food poisoning outbreaks and in many cases of illness in humans and animals and mortality in humans and animals.

Epidemiological data indicate that the cases of human or animal infections by *Salmonella* are may assume a dramatic dimension. In the European Union, the number of reported human cases is approximately 150 thousand each year, with a consistent tendency to decline during the last four years. In countries where basic health care cannot be delivered, the most dangerous clinical expression of a *Salmonella* infection, Typhoid fever, affects 16 million people per year, with almost 500,000 fatal cases, as estimated by the World Health Organization (Pang et al., 1998).

In fact, not all *Salmonella* isolates identified worldwide have such devastating consequences to human or animal health. Some of the *Salmonella* seroptypes or serovars are strictly adapted to primates (*Salmonella* Typhi, *Salmonella* Paratyphi, *Salmonella* Wien) being referred to as prototrophic for man; others are strictly adapted to some animal species (*Salmonella* Gallinarum-Pullorum, *Salmonella* Abortusovis, *Salmonella* Abortusequi), being referred to as prototrophic to animals; however, the vast majority of the serotypes are zoonotic, being able to infect both animals and humans. *Salmonella* varieties differentiation is based on its antigenic mosaic, in a complex combination of somatic (O), flagelar (H) and capsular (Vi) antigens. Serotyping according to the Kauffmann-White system, established in the middle of the last century, is still recognized as the reference method for discrimination of *Salmonella* varieties. Each combination of different antigens found in a particular *Salmonella* isolate (serotype or serovar) has a specific designation, following the international nomenclature based on one hundred years of scientific contributions, which sometimes originated peculiar designations (Popoff & Le Minor, 1997; Grimont & Weill, 2007).

Salmonella is an infectious and contagious bacterium that may be transmitted to humans, warm blood animals and reptiles by contaminated drinking water, by raw foods consumption, by direct contact with previously infected humans or animals and by iatrogenic accidents.

Food, in particular raw food, is the most common pathway for *Salmonella* infection, especially by zoonotic serotypes. Since the 1950's, zoonotic *Salmonella* became dominant in human salmonellosis cases. Virulence is strictly dependent on the serotype and it also varies with individual competences of each bacterial strain and with host susceptibility. These features explain why some serotypes have a higher prevalence in a particular host. In the last decades, two serotypes of zoonotic *Salmonella* showed a clear dominant incidence: *Salmonella* Typhimurium, firstly found in bovines, pigs, pigeons and secondarily in humans, while *Salmonella* Enteritidis, more common in poultry, but firstly found in humans, with the exception of Europe (EFSA, 2010b). These two serotypes may be discriminated more deeply using epidemiological markers like phage typing, molecular genotyping or other methodologies, including profiling for antimicrobial resistance phenotypes (R-type).

Salmonella Typhimurium is a somatic group B strain with the following antigenic formula: 4,[5],12:i:1,2. It has been early recognized as a serotype with a variable antigenic structure, like the lack of the somatic antigen [5], assuming, in this case, the designation of "variety Copenhagen" (very frequent in pigeons and bovines). The antigenic structure modulation of *Salmonella* Typhimurium may be mediated by plasmids, phages or proto-phages infections or segregations.

By the end of the 1980s, some isolates of a monophasic form of *Salmonella* Typhimurium - serotype I,4,[5],12:i:- gained epidemiological relevance, being more and more frequently referred to in the literature (Machado & Bernardo, 1990). The prevalence of this monophasic serotype has grown, presently being one of the most common *Salmonella* serotypes isolated from humans in several countries (Hopkins, et al., 2010) (Table 1). Other variants have also been found: variants lacking the 1st flagellar phase or lacking both (i and 1,2); and also variants without the somatic 1 antigen. The possibility that before the 1990s the scarce reports of *Salmonella* I,4,[5],12:i:- isolation, may reflect the difficulties in serotyping it being the isolates probably designated as *Salmonella* Typhimurium. At that time it was frequent to report some *Salmonella* serotypes as "Group B" or "untypable" (Switt et al., 2009).

Years of isolation	Country	Source
1986	Portugal	Chicken carcasses
1993	Thailand	Human
1997	Spain	Human
1991	Brazil	Human
1998	United States	Human
2000	Germany	Human, food, swine, cattle, broiler
2003	Italy	Human, swine
2005	UK	Human
2006	Luxembourg	Pork, pigs

Table 1. First time reports of *Salmonella* 4,[5],12:i: isolation (Adapted from EFSA, 2010b)

2. Occurrence of *Salmonella* I,4,[5],12:i:-

Cases of human infections with serovar I,4,[5],12:i:- have been related to severe illnesess. This serovar was responsible for an outbreak in New York City in 1998, in which 70% of the cases required hospitalization, being also associated with cases of systemic infections in Thailand and in Brazil (Switt et al., 2009).

Some foodborne outbreaks due to *Salmonella* I,4,[5],12:i:- have been reported in Europe. In 2006, Luxembourg signaled two outbreaks caused by a monophasic S. Typhimurium DT 193, corresponding to 133 human cases, 24 hospitalizations and one death (Mossong et al., 2007). Pork meat has been incriminated in these *Salmonella* cases (Mossong et al., 2007).

In Germany, the number of *Salmonella* I,4,[5],12:i:- related with human diseases has increased since 2000 (Hauser et al., 2010). Since 2006, the same monophasic variant of the multidrug-resistant *Salmonella* DT 193 strain has been associated with sporadic cases of salmonellosis, with increasing rates of hospitalization (Trupschuch et al., 2010). In 2008, the monophasic I,4,[5],12:i:- variant correspond to up to 42% of all S. Typhimurium isolates responsible for human salmonellosis (Hauser et al., 2010).

In France, official data suggest a gradual increase of *Salmonella* I,4,[5],12:i:- isolation rate in humans. After 2005, the frequency of this particular serovar raised from the eleventh to the third place (AFSSA, 2009) and a further significant increase was reported in the first five months of 2010 (Bone et al., 2010). In 2008, several outbreaks of *Salmonella* I,4,[5],12:i:- infections were identified in this country, including 13 family outbreaks, three collective infections and two hospital infections (AFSSA, 2009).

In Spain, the number of *Salmonella* I,4,[5],12:i:- illness cases has consistently increased since 1997, the year of the first report. Nowadays, the monophasic *Salmonella* is at the top five among the most frequently isolated *Salmonella* serovars in Spain (de la Torre et al., 2003). The epidemiological relevance of this serotype makes it a major cause of concern in Spain since the beginning of this Century (Echeita, et al., 1999; Guerra, et al., 2000).

In the UK, human infections by *Salmonella* I,4,[5],12:i:- began to be reported in 2005, when 47 cases occurred. In 2009 151 cases occured, representing an increase of more than 30%. Almost 30 % of the *Salmonella* I,4,[5],12:i:- isolates had a R-type ASSuT. In Scotland there was also an increase in the number of reports of *Salmonella* monophasic Group B cases.

Since 2008, sporadic and diffuse outbreaks related with ready to eat food have also been described in the UK linked to a DT 191A *Salmonella* I,4,[5],12:i:- strain, which is tetracycline-resistant (Peters et al., 2010). This strain is thought to have originated from infected frozen feeder mice imported into the UK for feeding exotic pets.

In Italy, *Salmonella* I,4,[5],12:i:- is one of the most frequent serotypes related to human cases of salmonellosis (Dionisi et al., 2009). R-type ASSuT represented 75% of the monophasic isolates identified in 2008 and 2009. Almost 50 % of the monophasic isolates were identified as *Salmonella* DT193 and 13% as *Salmonella* U302.

In The Netherlands, *S.* I,4,[5],12:i:- was related to human cases for the first time in 2004. After that, the number of cases has consistently grown, being the third most prevalent serotype responsible for human salmonellosis from 2005 to 2008 (Van Pelt et al., 2009).

There also many cases reported outside Europe. In the USA, *Salmonella* I,4,[5],12:i:- frequency in human infections has consistently increased from 2002, being now ranked in the top six. During 2007, some *Salmonella* I,4,[5],12:i:- outbreaks occurred in the USA, related to frozen chicken pies consumption and also to direct contact with turtles kept as pets (CDC, 2007a, 2007b).

Some cases have also been reported in Canada (Switt et al., 2009).

Salmonella I,4,[5],12:i:- has also been reported in Brazil quite early, in the 1970s. In São Paulo State, the occurrence of the strain in human infections was reported in the 1990s. Since then, the frequency of foodborne outbreaks and of extra-intestinal infections in humans promoted by this serovar showed a consistent tendency to increase (Tavechio et al., 2009).

In Thailand, *Salmonella* I,4,[5],12:i:- has been classified among the top five *Salmonella* serovars responsible for cases of foodborne salmonellosis (Amavisit, et al., 2005; Pornruangwong et al., 2008).

Human cases of infection with this particular serotype seem to be generally linked to raw meat. According to the EFSA zoonoses reports, in 2008 *Salmonella* I,4,[5],12:i:- has been related to 3.1% of *Salmonella* isolations in pig herds; in 2009, the same serotype has been found in 1.2% of the *Salmonella* positive bovine herds, 3.2% of positive pig herds and represented 1.4% of *Salmonella* isolations in poultry meat.

A particularly relevant feature of *Salmonella* I,4,[5],12:i:- is the fact that most virulent isolates exhibit a plasmid-mediated resistance to a wide range of antimicrobial compounds. Similar to its ancestral lineage - *Salmonella* Typhimurium DT104 - the monophasic strain I,4,[5],12:i:- frequently expresses a multiple resistance to ampicillin (A), streptomycin (S), sulphonamides (Su) and tetracyclines (T). This ASSuT antimicrobial resistance pattern is chromosomally-encoded (Hopkins et al., 2010).

The progressive increase of the incidence of this serotype lead some authors to consider *Salmonella* I,4,[5],12:i:- as a possible new pandemic strain (Hopkins, et al., 2010).

Data on the number of salmonellosis cases or outbreaks occurring in livestock due to *Salmonella* I,4,[5],12:i:- is not available. This subject needs to be further studied.

3. Characterization of monophasic *Salmonella enterica* subsp. *enterica* serovar I,4,[5],12:i:-

Serotyping divides *Salmonella* subspecies into subtypes, or serovars, based on the immunologic characterization of surface structures, such as O, H and in some cases Vi-antigens, through the use of polyvalent and monovalent antisera. The full antigenic pool of *Salmonella* [I,4,5,12:i:-] indicates that the somatic O-antigens expressed are I,4,[5],12. The underlined O factor 1 (1) means that this factor is determined by phage conversion, being present only if the culture is lysogenized by the corresponding converting phage. The factor 5 between square brackets ([5]) means that the antigen may be present or absent, not having a relation with phage conversion. So, in this serovar both factors (1 and 5) can be present or absent.

Most *Salmonella* strains are biphasic and express two serologically distinct flagellar antigens. The two antigens were historically designated as phases and the expression of two different phases is mediated at molecular level by an intricate mechanism unique to *Salmonella*. The regulation of phase 1 and phase 2 antigen expressions is under the control of the recombinase Hin. This recombinase facilitates the inversion of a promoter element so that it either (i) transcribes *flj*B (which encodes the phase 2 antigen FljB) and *flj*A (which encodes a repressor of *fli*C, the gene encoding the phase 1 antigen FliC) (Aldridge et al., 2006; Yamamoto & Kutsukake, 2006) or (ii) does not transcribe either of these genes. If the orientation of this promoter does not allow the transcription of *flj*B and *flj*A, the lack of repression of *fli*C transcription leads to the expression of phase 1 flagellar antigens.

Strains expressing both flagellar types are called biphasic. In contrast, strains defined as monophasic fail to express either phase 1 or phase 2 flagellar antigens. *S*. [I,4,5,12:i:-] possess only the phase 1 of the H-antigen "i" and lacks the second phase H antigen, encoded by *flj*B, which either is not present or contains mutation(s) affecting its expression. In 2007, Zamperini et al. screened *S*. 4,[5],12:i:- isolates for phase 1 and phase 2 antigen genes, *fli*C and *flj*B, and found that 100% of the isolates were positive for *fli*C, while 11% were positive for *flj*B. Approximately 89% of these isolates contained complete or partial deletions of the phase 2 flagellin gene, *flj*B, whereas 96% possessed the upstream gene, *hin*, which encodes the DNA invertase involved in "flipping" the *flj*B promoter.

Phage typing is a method also used for *Salmonella* typing based on the lysis of isolates with a panel of bacteriophages. Since this technique does not depend on the presence of the second phase H antigen, monophasic *Salmonella* reactions are performed with the same panel of phages used for *Salmonella* serovar Typhimurium (Echeita et al., 2001; Amavisit et al., 2005; Mossong et al., 2006). Thus, all phage types that have been recognized so far within monophasic *Salmonella* have also been found in *S*. serovar Typhimurium.

For example, the multidrug-resistant *Salmonella* 4,[5],12:i:- strain detected in Spain in 1997 was lysed by the *S*. Typhimurium phage 10 (Echeita et al., 2001). Phage type U302 was also detected among *S*. 4,[5],12:i:- isolates in other countries, such as Denmark (Ethelberg et al., 2004) and Italy (Dinosi et al., 2009). This phage type has been considered closely related to DT104 (Briggs & Fratamico, 1999).

However, *S.* 4,[5],12:i:– isolates have also been classified in other phage types linked to *S.* Typhimurium. In Germany, Hauser et al. (2010) analyzed *S.* [4,[5],12:i:– isolates obtained from different sources (human, swine and pork) and classified 70% of strains as DT193 and 19% as DT120. In another study, Hopkins et al., (2011) screened a large number of *S.* 4,[5],12:i:– strains from different countries (France, The Netherlands, England and Wales, Germany, Italy, Spain and Poland), obtained from similar sources as described by Hauser et al. (2010), and were able to identify 16 different phage types. However, the most commonly identified phage types were DT193, DT120 and RDNC ("Reaction Does Not Conform"). DT193 was the most common phage type identified in England and Wales, France, Germany, Spain and the Netherlands, while DT120 predominated in Italy and Poland. In other studies, *S.* 4,[5],12:i:– DT193 strains were also isolated from human cases of infection and/or pigs in United Kingdom, Luxembourg, United States and Spain (Hampton et al., 1995; Gebreyes & Altier, 2002; de la Torre et al., 2003; Mossong et al., 2006), while monophasic DT120 strains were identified in Italy (Dionisi et al., 2009).

According to serological characterization, it is difficult to identify the origin of the monophasic strains. This strain may be a new variant of the rare serovar Lagos (4,[5],12:i:–), or a new variant of the very common serovar Typhimurium [4,5,12:i:1,2], or even a new variant of other serovars with similar antigenic pools, such as *S.* Agama [4,12:i:1,6], *S.* Farsta [4,12:i:e,n,x], *S.* Tsevie [4,12:i:e,n,z15], *S.* Cloucester [1,5,12,27:i:l,w], *S.* Tumodi [1,4,12:i:z6] or as *S.* 4,5,27:i:z35, an unnamed serotype (Switt et al., 2009). However, large scale studies suggest that *S.* 4,[5],12:i:– is genetically related to Typhimurium [4,5,12:i:1,2], and is likely to have originated from a *S.* Typhimurium ancestor (Echeita et al., 2001; Zamperini et al., 2007). Monophasic *Salmonella* could have evolved by two distinct pathways. It could represent ancestral forms which did not acquire, though evolution, a second flagellar antigen or the required switching mechanism. Alternatively, it could originate as mutants of biphasic *Salmonella*, which have lost either the switching mechanism or the ability to express the second flagellar antigen (Burnens et al., 1996).

The atypical *fljB*-negative and multidrug-resistant *S.* 4,[5],12:i:– which emerged and spread in Spain in 1997 had a unique sequence specific for *S.* Typhimurium phage types DT104 and U302 and also an IS*200* fragment located in a Typhimurium serovar-specific location. Both facts strongly suggest that these strains are monophasic variants of *S.* Typhimurium (Echeita et al., 2001).

On other hand, *S.* 4,[5],12:i:– DT193 and DT120 strains were classified as monophasic variants of *S.* Typhimurium due to the presence of a Typhimurium-specific fragment of the malic acid dehydrogenase gene (Hopkins et al., 2011). However, these strains were negative for the DT104- and U302-specific region. Houser et al. (2010) indicated that phage type DT193 and DT120 isolates of both serovars presented genetic differences and represent different Pulsed-field Gel Electrophoresis (PFGE) clusters. Such differences seem to indicate that the *S.* Typhimurium phage type DT193 lineage was not a direct ancestor of the monophasic phage type DT193. In contrast, *S.* 4,[5],12:i:– phage type DT120 strains showed a higher genetic similarity with the *S. enterica* Typhimurium phage type DT120 strains, suggesting that this biphasic subtype was the recent common ancestor of the monophasic variant.

Different mutations and deletions have been associated with the lack of phase 2 flagella expression in *S.* 4,[5],12:i:– isolates. Specifically, some Spanish *S.* 4,[5],12:i:– isolates appear to be characterized by the deletion of a large fragment, that included *fljB*, *hin*, and a DNA invertase essential for *fljB* expression (Garaizar et al., 2002). Most USA isolates characterized so far also present deletions that eliminate *fljB* but maintain *hin* (Zamperini et al., 2007; Soyer et al., 2009). These genetic differences among American and Spanish *S.* 4,[5],12:i:– isolates have also been made evident by PFGE typing (Soyer et al., 2009). Genetic data indicate that American and Spanish isolates represent different clonal groups with distinct genome deletion patterns. This is consistent with the observation that most Spanish *S.* 4,[5],12:i:– isolates are phage type U302 (Echeita et al., 2001). Moreover, PFGE and Multiple-Locus Variable Number Tandem Repeat Analysis (MLVA) techniques showed that Spanish phage type U302 strains seem to be more homogeneous than groups constituted by isolates from other countries (Soyer et al., 2009), supporting a clonal origin. Soyer et al. (2009) suggested that Spanish *Salmonella* 4,[5],12:i:– strains might have emerged from a multidrug-resistant *S.* Typhimurium strain, while American *S.* 4,[5],12:i:– strains might have emerged from a non-drug-resistant *S.* Typhimurium strain, through independent events.

Strains belonging to *S.* 4,[5],12:i:– that have been recently implicated in infections both in humans and farm animals have been further typed using genetic techniques. Guerra et al. (2000) studies showed a high genetic homogeneity among 16 Spanish *S.* 4,[5],12:i:– isolates using techniques such as ribotyping, RAPD (Random Amplified Polymorphic DNA analysis) and plasmid profiling. However, the use of PFGE techniques could originate degrees of heterogeneity that may range from moderate to high, even when applied to strains from a single country (Agasan et al., 2002; de la Torre et al., 2003; Zamperini et al., 2007; Soyer et al., 2009). For example, at least 13 different *Xba*I PFGE types were found among 32 *S.* 4,[5],12:i:– isolates from Georgia (Zamperini et al., 2007), 44 different *Xba*I PFGE types were detected among 148 isolates from Germany (Hauser et al., 2010) and at least 11 *Xba*I PFGE types were found among 23 Spanish *S.* 4,[5],12:i:– isolates (de la Torre et al., 2003). Despite their heterogeneity, *S.* 4,[5],12:i:– strains have been reported to be less heterogenic than *S.* Typhimurium strains (Guerra et al., 2000; Agasan et al., 2002; Soyer et al., 2009; Hauser et al., 2010).

Studies also have showed the occurence of common genetic profiles among *S.* 4,[5],12:i:– isolates obtained from different sources and countries. Zamperini et al. (2007) studies revealed the same PFGE profile among poultry and bovine *S.* 4,[5],12:i:– isolates. In 2011, Hopkins et al. compared isolates from humans, pigs and pork using PFGE and detected several prevalent genetic profiles common to these three sources (StYMXB.0131, STYMXB.0083, STYMXB.0079, STYMXB.0010, STYMXB.0022). Moreover, these profiles were detected in isolates from different countries. One PFGE profile, STYMXB.0010, was identified in isolates obtained in all the countries surveilled in the referred study. However, the authors also found some country-specific differences in the distribution of PFGE patterns. For example, nine of the 12 STYMXB.0079 strains originated from Italy, three of the five Polish strains were STYMXB.0010 and six of 10 strains from the Netherlands were STYMXB.0131. The STYMXB.0079 profile was also the predominant one among 146 human *S.* 4,[5],12:i:– isolates obtained in Italy by Dionisi et al. (2009).

Typing of monophasic strains using molecular techniques, such as PFGE, have showed that these strains differ from *S.* Lagos strains (Soyer et al., 2009). Data also showed the occurrence of some profiles common to *S.* Typhimurium. Zamperini et al. (2007) examined isolates of *S.* 4,[5],12:i:– and *S.* Typhimurium collected from animal sources that presented the same PFGE profile. In another study, Agasan et al. (2002) compared the PFGE profile of S. 4,[5],12:i:– strains found in humans in New York City to the profile of S. Typhimurium isolates, including S. serovar Typhimurium DTI04, and found that S. 4,[5],12:i:– isolates were related to some of the S. Typhimurium isolates examined. Amavisit et al. (2005) compared the PFGE profiles of human isolates identified as S. Typhimurium DTI04, Typhimurium U302 and 4,[5],12:i:–, showing that four of the S. 4,[5],12:i:– isolates presented the same or similar profiles as S. Typhimurium phage type U302.

In another study, Alcaine et al. (2006) used Multilocus Sequence Typing (MLST) to show that ST6 type comprises not only bovine and human S. 4,[5],12:i:– isolates but also Typhimurium isolates obtained in the United States. ST6 was unique to S. Typhimurium and 4,[5],12:i:–, which supports the initial findings based on characterization of Spanish isolates (de la Torre et al., 2003), that S. [4,5,12:i:-] may have emerged from a S. Typhimurium ancestor. A MLST technique based on four genes applied to American and Spanish isolates belonging to S. 4,[5],12:i:– and to S. Typhimurium classified the vast majority of isolates as ST1 (Soyer et al., 2009). Similar results were obtained with other molecular fingerprinting techniques e.g. RAPD analysis, plasmid profiling (Sala, 2002; de la Torre et al., 2003), MLST (Alcaine et al., 2006) and MLVA/Variable Number of Tandem Repeats (VNTR) typing (Laorden et al., 2009; Torpdahl et al., 2009; Hauser et al., 2010; Hopkins et al., 2011). All these studies lead to the conclusion that S. 4,[5],12:i:– isolates belong to a single genetic lineage or clone and seem closely related to S. Typhimurium (Zamperini et al., 2007; Dionisi et al., 2009; Hopkins et al., 2011).

Overall, the genomic characterization of S. 4,[5],12:i:– isolates suggests that this serovar is likely to gather several clones or strains that have independently emerged from S. Typhimurium during the last two decades, and have changed through multiple independent events involving different clonal groups (Garaizar et al., 2002; Laorden et al., 2009; Laorden et al., 2010). Although the driver for this evolution remains to be enlightened for many epidemic strains antimicrobial resistance may be implicated (Zaidi et al., 2007; Bailey et al., 2010).

4. Antimicrobial resistance traits of *Salmonella* I,4,[5],12:i:-

The characterization of zoonotic bacteria virulence factors, including the presence of antimicrobial resistance traits, is of major importance for assuring the safeguard of health in the wider concept of "one health". The dissemination of antimicrobial resistant bacteria is a well-recognized hazard for public and animal health.

Several *Salmonella* serovars are frequently related to human and animal diseases, and this genus is recognized worldwide as a major foodborne pathogen. Gastroenteritis due to *Salmonella* is usually characterized by mild to moderate self-limiting symptoms, such as diarrhea, abdominal cramps, vomiting and fever. However, some strains are responsible for severe infections, such as septicemia, osteomyelitis, pneumonia, and meningitis that occur, especially in children and in elderly and immunocompromised individuals (Folley and Lynne, 2008a).

Generally, salmonellosis cases caused by *Salmonella* I,4,[5],12:i:- strains are severe, requiring hospitalization (EFSA, 2010b). The control of severe infections requires antimicrobial therapy, generally with fluoroquinolones or ceftriaxone, administrated to children in order to avoid the cartilage damage frequently associated with fluoroquinolone therapy (Folley and Lynne, 2008a). Therefore, *Salmonella* represents a bacterial genus of special concern regarding antimicrobial resistance dissemination.

This serotypes' resistance profiles may vary, worldwide, from 100% susceptible to multidrug resistance. Although *S.* Typhimurium resistance levels have been decreasing in several European countries, the incidence of resistant *S.* Typhimurium I,4,[5],12:i:- strains seems to be escalating (Switt et al., 2009). There are only a few studies available on antimicrobial resistance traits and genes present in antimicrobial drug–resistant *Salmonella* serotype 4,[5],12:i:- isolates, which have identified some specific resistance genes and genetic mechanisms. The limited data available still hasn't allowed researchers to identify the common ancestor responsible for the emergence of 4,[5],12:i:- isolates with a multidrug resistance pattern (MDR), information essential to understand resistance evolution and dissemination (Switt et al., 2009).

Antimicrobial resistance in *Salmonella* spp. may be due to several resistance determinants that can be located either in the bacterial chromosome or in plasmids (Folley and Lynne, 2008a; Switt et al., 2009). These genetic determinants can be responsible for the expression of intrinsic resistant mechanisms, related to the production of ß -lactamases, to the modification of the antimicrobial compound by bacterial enzymes, to the variation of bacterial permeability, to the presence of efflux pumps or to the modification of target receptors (Folley and Lynne, 2008a).

Antimicrobial resistance may also result from the expression of acquired resistance mechanisms, emerging through the occurrence of point mutations in chromosomal genes or the acquisition of mobile elements such as plasmids, transposons, and genomic islands (Switt et al., 2009). The transfer of resistance determinants may occur directly from the same or different bacterial species/genera, or indirectly through the environment (Folley and Lynne 2008a; EFSA, 2010b). Intestinal microbiota from humans and animals is often exposed to antimicrobial compounds of different classes, concentrations and exposure frequencies, used for therapy, prophylaxis or methaphylaxis. This exposure may derive from food/feed products or from the environment (Martins da Costa et al., 2007). Emergence, selection and dissemination of antimicrobial resistant bacteria are still mainly attributed to the selective pressure of antibiotic misuse and abuse (Monroe & Polk, 2000; Sayah et al., 2005), so intestinal bacteria can became resistant to some antimicrobial compounds, and therefore transmit these resistant traits to *Salmonella*, which occupies the same ecological niche.

The presence of one or the combination of several of the above mentioned mechanisms may also confer a MDR profile to bacteria. These MDR profiles may comprise major antimicrobial compounds, hampering the treatment of severe *Salmonella* infections (EFSA, 2010b).

In 1997, MDR *Salmonella* 4,[5],12:i:- isolates were identified for the first time in Spain (Guerra et al., 2001). The most frequent MDR pattern is the ASSuT tetraresistance pattern, isolated from 30% of the human infection cases in the last 5 years and also from farm animals (Lucarelli et al., 2010; EFSA, 2010b). This pattern emerged in Italy during the 2000s, and has already been identified in Denmark, the United Kingdom, the United States, Spain, France, and the Czech Republic (Lucarelli et al., 2010). Genes responsible for this MDR phenotype are present in a chromosomal resistance island that usually includes the *bla*TEM, strA-strB, sul2 and tet(B) genes (Hauser et al., 2010; Lucarelli et al., 2010), having some strains additional resistances (Lucarelli et al., 2010; EFSA, 2010b).

Other multirresistant patterns identified in 4,[5],12:i:– isolates worldwide are the ACKGSuTm (showing resistance to ampicillin, chloramphenicol, kanamycin, gentamicin, sulfamethoxazole and trimethoprim) and ACKGSuTm with additional resistance to nalidixic acid patterns, found in Thailand (Switt et al., 2009); the ACSuGSTTm (showing resistance to ampicillin, chloramphenicol, sulfamethoxazole, gentamicin, streptomycin and tetracycline) and ACGSSuTSTm patterns, found in Spain (Echeita et al., 1999); the ACSSuT pattern, found in the United States (Agasan et al., 2002; Switt et al., 2009); and the ACSSpSuT pattern, found in the United Kingdom and other countries (Lucarelli et al., 2010). The isolation of multiresistant isolates was also described in Brazil (Switt et al., 2009) and Germany (Hauser et al., 2010).

The MDR phenotypes include 4,[5],12:i:– strains harboring class 1 integrons or large resistant plasmids, resistant to ampicillin, chloramphenicol, gentamicin, streptomycin, sulfamethoxazole, tetracyclines and trimethoprim. These resistance traits are mainly due to the expression of blaTEM-1, which codes for broad spectrum b-lactamases responsible for resistance to penicillin and amino-penicillins; of blaCTX-M-1, which codes for extended-spectrum ß-lactamases; of cmlA1, which codes for an efflux pump responsible for chloramphenicol resistance; of aac(3)-IV and aadA2, which code for enzymes that modify gentamycin and streptomycin active sites, impairing the action of these drugs; of aadA1, sul1 and sul2, which code for enzymes responsible for resistance to sulfonamides; of sul3 and tet(A), which code for an efflux pump mechanism responsible for tetracycline resistance; and of dfrA12, which codes for an enzyme responsible for resistance to trimethoprim (Folley and Lynne, 2008a; Guerra et al., 2001; Switt et al., 2009).

It is important to refer that, despite the road book aiming at controlling antimicrobial use and abuse, antimicrobial resistance remains a worldwide problem for both human and veterinary medicine. In this context, the boundaries between human and animal health, as well as between living organisms and the environment are insubstantial. Besides data from clinical studies, resistant bacteria have been described from a variety of environmental sources, including domestic sewage, drinking water, rivers, and lakes (Sayah et al., 2005).

5. *Salmonella* virulence factors

Salmonella enterica includes many serovars that cause disease in avian and mammalian hosts (Eswarappa et al., 2008). Also, *Salmonella* sp. is one of the most frequent bacterial food-borne pathogens affecting humans. In both animal and human hosts, infections may be present in a variety of presentations, from asymptomatic colonization to inflammatory diarrhoea or typhoid fever, depending on serovar- and host-specific factors. Colonization of reservoir hosts often occurs in the absence of clinical signs; however, some *S. enterica* serovars threaten animal health due to their ability to cause acute enteritis or to translocate from the intestine to other organs, causing fever and septicaemia (Stevens, 2009). Also, while certain serovars of *S. enterica* are ubiquitary and cause disease in humans and in a variety of animals, other serovars are highly restricted to a specific host (Hensel, 2004). For example, ubiquitous serovars such as Typhimurium and Enteritidis tend to produce an acute but self-limiting enteritis in a wide range of hosts, whereas host-specific serovars are associated with severe systemic disease that may not involve diarrhoea, usually affecting healthy adults of a single species (e.g. *S.* Typhi in humans, *S.* Gallinarum in poultry) (Stevens, 2009).

Differences in virulence among *Salmonella* serovars and variations in the evolution of *Salmonella* spp. infections in several host species have been attributed to the acquisition and expression of virulence genes (Zhao, 2001). *Salmonella* spp. virulence requires the coordinated expression of complex arrays of virulence factors that allow the bacterium to evade the host's immune system. All *Salmonella* serotypes share the ability to invade the host by inducing their own uptake into the intestinal epithelial cells. In addition, *Salmonella* serotypes associated with gastroenteritis trigger an intestinal inflammatory and secretory response, whereas serotypes that cause enteric fever give raise to systemic infections through their ability to survive and replicate in mononuclear phagocytes (Ohl & Miller, 2001).

Many virulence phenotypes of *Salmonella enterica* are encoded by genes located in distinct chromosome regions, organized in 12 pathogenicity islands (Bhunia, 2008; Eswarappa et al., 2008; Saroj et al., 2008). These gene clusters, known as *Salmonella* pathogenicity islands (SPIs), are thought to be acquired by horizontal gene transfer. They present a G-C content that differs from the remaining chromosome, suggesting acquisition by horizontal transfer. While some SPIs are conserved throughout the genus, others are specific for certain serovars (Amavisit et al., 2003; Bhunia, 2008; Eswarappa et al., 2008). According to Saroj et al., (2008), pathogenicity islands can be transferred between bacteria of different genera, leading to an accumulation of different virulence mechanisms in some strains. Therefore, the occurrence of SPIs varies between serovars and strains (Hensel, 2004). Pathogenicity islands often contain multiple genes functionally related, and required for the expression of a specific virulence phenotype, which suggests that the acquisition of a pathogenicity island during evolution may in one "quantum leap" open up new host niches for the pathogen (Ohl & Miller, 2001; Eswarappa et al., 2008). According to Bhunia (2008), the virulence genes responsible for invasion, survival, and extraintestinal spread are distributed in the *Salmonella* pathogenicity islands. For instance, the virulence genes that are involved in the intestinal phase of infection are located in SPI-1 and SPI-2. Many pathogenicity islands, including SPI-1 and SPI-2, encode specialized devices for the delivery of virulence proteins into host cells, termed type III secretion systems (TTSSs) (Eswarappa et al., 2008). The remaining SPIs are required for causing systemic infections, intracellular survival, fimbrial expression, antibiotic resistance, and Mg2+ and iron uptake (Bhunia, 2008).

Besides the SPIs, some virulence factors can be encoded in virulence plasmids. Six serovars (Typhimurium, Gallinarum, Gallinarum biovar Pullorum, Enteritidis, Dublin, Choleraesuis and Abortusovis) typically harbor virulence plasmids of 60-95 kb that contain the *spv* locus, which holds some of the genes that are involved in intracellular survival and multiplication of this facultative intracellular pathogen (Tierrez & Garcia-del Portillo, 2005). The typical virulence plasmid of *S*. Typhimurium (pSLT90), is about 90-95 kb, and belongs to the FII incompatibility group.

Regarding the monophasic *S*. Typhimurium, this serotype has only recently emerged, but it comprises a wide variety of different strains (Soyer et al., 2008). For that, consistent data on virulence mechanisms are limited. Nevertheless, several studies have already shown that not only *Salmonella* serotype 4,[5],12:i:– isolates are genetically and phenotypically closely related to *Salmonella* serotype Typhimurium (Agasan et al., 2002; Amavisit et al., 2005; de la Torre et al. 2003; Delgado et al., 2006; Echeita et al., 2001; Zamperini et al., 2007) but also, virulence genes of monophasic *S*. Typhimurium and their variability are identical to those found in *S*. Typhimurium (Garaizar et al., 2002; Hauser et al., 2009; Soyer et al., 2009; Hauser et al., 2010). For example, studies developed by del Cerro et al. (2003) and Guerra et al. (2000), demonstrated that strains of monophasic *S.* Typhimurium presented an homology regarding virulence plasmid genes *spvC*, *invE* and *invA* invasion genes, *stn* enterotoxin genes, *slyA* cytolysin genes and genes associated with survival within macrophages (*pho*), when compared to those typically found in *S.* Typhimurium

For all these reasons, it should be noted that, presently, most of the knowledge on SPIs and other *Salmonella* virulence genes of monophasic *S.* Typhimurium is based on observations made in serovar Typhimurium. This serovar is considered a model organism for genetic studies, and a wide variety of classical and molecular tools are available for the identification and characterization of potential *Salmonella* virulence genes.

5.1. *Salmonella* Pathogenicity Islands

As above referred, there are at least twelve chromosomally-encoded *Salmonella* pathogenicity islands (SPIs) (Table 2), as follows:

- SPI-1 is a 43-kb chromosomal locus that was acquired by horizontal gene transfer from other pathogenic bacteria during evolution. It contains 31 genes with a major role in the invasion of host cells and induction of macrophage apoptosis. It also encodes components of the Type III secretion system (TTSS) designated as the Inv/Spa-Type III secretion apparatus that includes the secretion apparatus components, effectors, chaperones, and regulator (Amavisit et al., 2003; Bhunia, 2008; Eswarappa et al., 2008). The major genes present in SPI-1 are *invA*, *invB*, *invC*, *invF*, *invG*, *hilA*, *sipA*, *sipC*, *sipD*, *spar*, *orgA*, *sopB*, and *sopE*. *invABCD* genes, responsible for the expression of several invasion factors that promote bacterial attachment and invasion of M-cells, allowing them to cross the epithelial barrier which is the preferential route of *Salmonella* translocation. For example, InvA is an inner membrane protein involved in the formation of a channel through which polypeptides are exported. InvH and HilD are accessory proteins involved in *Salmonella* adhesion. InvG is an outer membrane protein of the TTSS that plays a critical role in bacterial uptake and protein secretion.

    There are two kinds of effector proteins secreted by the TTSS. One subclass consists of InvJ and SpaO, which are involved in the protein secretion through the TTSS. The other subclass modulates host cytoskeleton and induces its uptake. SipB and SipC are the major proteins, which interact with host cytoskeletal proteins to promote *Salmonella* uptake. Inv/Spa are also responsible for macrophage apoptosis. SipA is an actin-binding protein. SopB is an inositol phosphate phosphatase and SopE activates GTP-binding proteins. HilA is the central transcriptional regulator of genes located on SPI-1 (Bhunia, 2008).

- SPI-2 is a 40-kb segment that encodes for 32 genes, only present in members of *S. enterica*, and other type III secretion systems involved in systemic pathogenesis (Amavisit et al., 2003; Eswarappa et al., 2008; Bhunia, 2008). The gene products are essential for systemic infection and mediate bacterial replication, rather than survival within host macrophages (Bhunia, 2008). The majority of these genes are expressed during bacterial growth inside the host-cells. SPI-2 carries genes for Spi/Ssa and TTSS apparatus, i.e., SpiC, which inhibits the fusion between the *Salmonella*-containing phagosome and the lysosome (Bhunia, 2008).

- Type III Secretion Systems are expressed by many bacterial pathogens to deliver virulence factors to the host cell and to interfere with or subvert normal host cell signaling pathways (Marcus et al., 2000). The TTSS structural genes (including *inv*G, *prg*H and *prg*K) encode proteins that may form a needle-like structure and are responsible for contact dependent secretion or for the delivery of virulence proteins to host cells (Zhao, 2001; Bhunia, 2008). This needle-like organelle located in the bacterial periphery has four parts: a needle, outer rings, neck, and inner rings. The needle is constituted by PrgI and a putative inner rod protein, PrgJ; the outer rings structure by InvG; the neck by PrgK; and the base by PrgH that forms the inner rings. The inner membrane components include InvC, InvA, SpaP, SpaQ, SpaR, and SpaS proteins (Bhunia, 2008). When *Salmonella* adheres to a target cell, this needle-like structure is assumed to form a channel with its base anchored in the cell wall and its tip puncturing the membrane of the host cell. Through this channel, *Salmonella* effectors proteins such as SipC, SipA, SopE/E2, and SopB, are injected into the host cell cytoplasm, promoting actin polymerization and membrane remodelling which allows the active uptake of bacteria by the host cell (Zhao, Y., 2001).

- SPI-3 is a 17-kb locus conserved between *S. enterica* serovar Typhi and Typhimurium that is also found in *S. bongori*, being variable in other serovars. SPI-3 harbors 10 genes, including the *mgt*CB operon, which is regulated by PhoPQ and is required for intra-macrophage survival and virulence and for magnesium uptake under low magnesium concentrations (Amavisit et al., 2003; Bhunia, 2008; Eswarappa et al., 2008). PhoQ is a sensor and PhoP is a transcriptional activator that expresses different genes that are required for bacterial survival inside the macrophage, as well as in various stressing environments including carbon and nitrogen starvation, low pH, low O2 levels, and the presence of defensins. In addition, PhoP regulates genes such as *spi*C and *tass*C that prevent lysosome fusion with the *Salmonella*-containing vacuole. PhoQ regulon activates *pags* genes that are essential for adaptation during the intracellular life cycle (Bhunia, 2008).

Salmonella present in the subcellular lamina propria are either engulfed by the macrophages or by the dendritic cells, which allows its extraintestinal dissemination. The survival of *Salmonella* within macrophages is generally considered to be essential for the translocation of bacteria from the gut-associated lymphoid tissue to the mesenteric lymph nodes and from there to the liver and spleen.

- SPI-4 is a 27-kb locus located next to a putative tRNA gene, containing 18 genes. It is thought to encode genes for the Type I secretion system and is suspected to be required for intramacrophage survival (Amavisit et al., 2003; Bhunia, 2008).
- SPI-5 is a 7.6-kb region and encodes six genes. It appears that SPI-5 encodes effector proteins for TTSS. SopB, which is translocated by TTSS, is an inositol phosphatase involved in triggering fluid secretion responsible for diarrhea. Thus, it is believed that SPI-5 is possibly responsible for enteric infections (Bhunia, 2008; Eswarappa et al., 2008).
- SPI-6 is a 59-kb locus present in both serovars Typhi and Typhimurium. It contains the *saf* gene cluster responsible for fimbriae development, *pag*N responsible for invasion traits, and several genes with unknown function (Bhunia, 2008).

In many studies, bacterial motility was found to be essential for adherence or invasion. In many systems, flagella provide the driving force that enable the bacteria to penetrate the host mucus layer and reach the host cell surface more rapidly (Zhao, 2001). *Salmonella* expresses different types of fimbriae that promote adhesion to M-cells and colonization of intestinal epithelial cells. Type I fimbriae (Fim) binds to α-d-mannose receptor in the host cell; long polar fimbriae (Lpf) bind to cells located in the Peyer's patch; and plasmid-encoded fimbriae (Pef) and curli, thin aggregative fimbriae, aid in bacterial adhesion to intestinal epithelial cells. Curli helps bacteria to autoaggregate, which enhances survival in the presence of stomach acid or biocides (Bhunia, 2008).

- SPI-7 or Major Pathogenicity Island (MPI) is a 133-kb locus specific for serovar Typhi, Dublin, and Paratyphi. Its genes encode for Vi antigen, a capsular polysaccharide that illicits high fever in typhoid fever infections. SPI-7 also carries the *pil* gene cluster responsible for type IV pili synthesis and the gene that encodes for the SopE effector protein of TTSS (Bhunia, 2008).

- SPI-8 is a 6.8-kb locus that appears to be specific for serovar Typhi. It carries genes for putative bacteriocin biosynthesis but its functional traits have not been fully investigated (Bhunia, 2008).

- SPI-9 is a locus of approximately 16-kb that carries genes for type I secretion system and a large putative RTX (repeat in toxin)-like toxin (Bhunia, 2008). SPI-9 is present in *S.* Typhi, and also as a pseudogene in *S.* Typhimurium (EFSA, 2010)

- SPI-10 is a 32.8-kb locus found in serovars Typhi and Enteritidis. It contains genes that encode for Sef fimbriae (Bhunia, 2008).

- *Salmonella* Genomic Island 1 is a 43-kDa locus that contains genes responsible for antimicrobial resistance. It was identified in *S.* Typhimurium DT104, Paratyphi and Agona, which are resistant to multiple antibiotics. The DT104 strain has been implicated in outbreaks worldwide. It includes genes responsible for five antimicrobial resistance phenotypes (ampicillin, chloramphenicol, streptomycin, sulphonamides, and tetracycline) that are clustered in a multidrug resistance region and are composed of two integrons (Bhunia, 2008).

- High Pathogenicity Island (HPI) contains genes responsible for siderophore biosynthesis, required for iron uptake. The HPI is found in *S. enterica* (Bhunia, 2008).

Islands	Salmonella serovars	Length Kb)	Functions
SPI - 1	*S. enterica* and *S. bongori*	43	TTSS, invasion of host cells
SPI - 2	*S. enterica*	40	TTSS, systemic infection
SPI - 3	*S. enterica* and *S. bongori*	17	Mg2+ uptake, macrophage survival
SPI - 4	*S. enterica* and *S. bongori*	27	Macrophage survival
SPI - 5	*S. enterica* and *S. bongori*	7.6	Enteropathogeniticity
SPI - 6	*S. enterica* subsp. *enterica*	59	Fimbriae
SPI - 7	*S.* Typhi, *S.* Dublin, *S.* Paratphy	133	Vi antigen
SPI - 8	*S.* Typhi,	6.8	Unknown; putative bacteriocin biosynthesis
SPI - 9	*S.* Typhy	16.3	Type I secretion system and RTX – like toxin
SPI - 10	*S.* Typhi, *S.* Enteritidis	32.8	Sef fimbriae
SGI - 1	*S.* Typhimurium (DT104), *S.* Partyphi, *S.* Agona	43	Antibiotic resistance genes
HPI	*S. enterica* subsp. IIIa, IIIb, IV	?	High affinity iron uptake

Table 2. Main properties and functions of *Salmonella* pathogenicity islands (SPI) (Adapted from Hensel, 2004 and Bhunia, 2008)

Presently, there are over 30 *Salmonella* specific genes that have been used as targets for PCR (Polymerase Chain Reaction) to detect and characterize *Salmonella*. These include *inv*A gene sequences that are highly conserved among all *Salmonella* serotypes, other gene sequences also present throughout the genus, and fimbriae protein-encoding genes and antibiotic resistance genes (Table 3).

Gene Description	Description
invA	Triggers internalization required for invasion of deep tissue cells
InvE/A	Invase proteins
phoP/Q	Intramacrophage survival and enhanced bile resistance
stnB	*Salmonella* enterotoxin gene
irob	Iron regulation
slyA	Salmolysin
hin/H2	Flagellar phase variation
afgA	Thin aggregative fimbriae
fimC	Pathogen related fimbrae gene of *S. enterica*
sefA	Major subunit fimbrial protein of serotype Enterica strains
pefA	Fimbrial virulence gene of *S*. Typhimurium
spvA	Virulence plasmid region
spvB	Virulence plasmid region
spvC	Virulence plasmid region that interacts with the host immune system and is responsible for an increased growth rate in host cells
rep-FIIA	Plasmid incompatibility group
sprC	Virulence gene
sipB–sipC	Junction of virulence genes *sipB–sipC*
himA	Encodes a binding protein
his	*Salmonella* genus specific histidine transport operon
prot6e	Virulence plasmid region specific for *S*. Enteritidis
ST M3357	Regulatory protein whose start codon sequence determines the DT phenotype exhibiting enhanced virulence

Table 3. Genes Used for the PCR Identification of *Salmonella* spp. (Adapted from Levin, 2010)

6. Conclusions

Salmonella spp. is one of the major foodborne pathogen responsible for outbreaks worldwide (EECDC, EFSA, 2009; Switt et al., 2009), being estimated to be the main pathogen responsible for foodborne mortality in the United States (Mead et al., 1999). This bacterial genus includes 2,500 identified serotypes, distributed between 2 species: *Salmonella* enterica and *Salmonella* bongori (Foley and Lynne, 2008a). The emergence of new pathogenic strains and serotypes has been described (EFSA, 2010b; Hauser et al., 2010). Due to their increased virulence, these strains can rapidly spread among production animals and humans, representing a major public health issue (EFSA, 2010b; Hauser et al., 2010). In the mid-1990s the emergence of *Salmonella enterica* subsp. *enterica* serotype I,4,[5],12:i:-, a monophasic variant of *Salmonella* Typhimurium, has been reported in Europe (Foley et al., 2008b; Hauser et al., 2010; Switt et al., 2009). Nowadays it seems to be one of the major serotypes responsible for human salmonellosis cases worldwide (EECDC, EFSA, 2009; Switt et al., 2009). It has also been isolated from several animal species, such as poultry, cattle, swine, and turtles, and also from food products, such as poultry and pork products.

In 2010, the European Food Safety Authority (EFSA) Panel on Biological Hazards (BIOHAZ) published a Scientific Opinion alerting for the increasing number of outbreaks in the European Union member states promoted by *Salmonella* Typhimurium-like" strains. The Panel has recommended that these strains should be further typed and characterized, particularly in terms of antimicrobial resistance (EFSA, 2010b).

Studies aiming at fully characterizing the monophasic variants of *Salmonella* Typhimurium-like strains (4,[5],12:i:–) isolated from different sources, such as food products, animals and the environment, in terms of molecular typing, antimicrobial resistance, virulence traits and immune response modulation, are extremely relevant. Data provided by such studies will have repercussions in preventive and therapeutic strategies, both in human and veterinary medicine.

7. Acknowledgements

This work was supported by CIISA ("Centro de Investigação Interdisciplinar em Sanidade Animal") from the Faculty of Veterinary Medicine, Lisbon. Manuela Oliveira is a FCT ("Fundação para a Ciência e a Tecnologia") funded scientist from the program "Ciência 2007".

8. References

AFSSA (2009). Opinion of the French Food Safety Agency concerning two draft amendments to Orders for controlling salmonellae in the *Gallus gallus* species. 2009-SA-0182, pp-22.

Agasan, A., Kornblum, J., Williams, G., Pratt, C.C., Fleckenstein, P., Wong, M., Ramon, A. (2002). Profile of *Salmonella enterica* subsp. *enterica* (subspecies I) serotype 4,5,12:i:-strains causing food-borne infections in New York City. *Journal of Clinical Microbiology*, 40, 6, (Jun 2002), pp. 1924-1929, ISSN 0095-1137.

Alcaine, S.D., Soyer, Y., Warnick, L.D., Su, W.L., Sukhnanand, S., Richards, J, Fortes, E.D., McDonough, P., Root, T.P., Dumas, N.B., Grohn, Y., Wiedmann. M. (2006). Multilocus sequence typing supports the hypothesis that cow- and human-associated *Salmonella* isolates represent distinct and overlapping populations. *Applied and Environmental Microbiology*, 72, 12, (Dec 2006), pp. 7575-7585, ISSN 0099-2240.

Aldridge, P.D., Wu, C., Gnerer, J., Karlinsey, J.E., Hughes, K.T., Sachs, M.S. (2006). Regulatory protein that inhibits both synthesis and use of the target protein controls flagellar phase variation in *Salmonella enterica*. *Proceedings of the National Academy of Sciences of The United States of America*, 103, 30, (Jul 2006), pp. 11340-11345, ISSN 0027-8424.

Amavisit, P., Lightfoot, D., Browning, G.F., Markham, P.F. (2003). Variation between pathogenic serovars within *Salmonella* Pathogenicity Islands. *Journal of Bacteriology*, 185, 12, (Jun 2003), pp. 3624-3635, ISSN 0021-9193.

Amavisit, P., Boonyawiwat ,W., Bangtrakulnont, A. (2005). Characterization of *Salmonella* enteric serovar typhimurium and monophasic *Salmonella* serovar I,4,[5],12 : i : - isolates in Thailand. *Journal of Clinical Microbiology*, 43, 6, (Jun 2005), pp. 2736-2740, ISSN 0095-1137.

Antunes, P., Mourão, J., Freitas, A., Peixe, L. (2010). Population structure of multidrug-resistant nontyphoidal *Salmonella enterica* isolates from Portugal. *Clinical Microbiology and Infection*, 16, S371-372, ISSN 1469-0691.

Bailey, A.M., Ivens, A., Kingsley, R., Cottell, J.L., Wain, J., Piddock, L.J. (2010). RamA, a member of the AraC/XylS family, influences both virulence and efflux in *Salmonella enterica* serovar Typhimurium. *Journal of Bacteriology*, 192, 6, (Mar 2010), pp. 1607-1616, ISSN 0021-9193.

Barone, D.L., Dal Vecchio, A., Pellissier, N., Viganò, A., Romani, C., Pontello, M. (2008). Emergence of *Salmonella* Typhimurium monophasic serovar: determinants of antimicrobial resistance in porcine and human strains. *Annali di igiene: medicina preventive e di comunità*, 20, 3, (May-Jun 2008), pp. 199-209, ISSN 11209135.

Bhunia, A.K. (2008). *Salmonella enterica. Foodborne Microbial Pathogens: Mechanisms and Pathogenesis*. Springer, pp 201-216, ISBN: 038774536X, USA.

Bone, A., Noel, H., Le Hello, S., Pihier, N., Danan, C., Raguenaud, M.E., Salah, S., Bellali, H., Vaillant, V., Weill, F.X., Jourdan-da Silva, N. (2010). Nationwide outbreak of *Salmonella enterica* serotype 4,12:i:- infections in France, linked to dried pork sausage. Euro Surveillance, 15, 24, (Mar-May 2010), pp. 1-3.

Briggs, C.E., Fratamico, P. (1999). Molecular characterization of an antibiotic resistance gene cluster of *Salmonella* typhimurium DT104. *Antimicrobial Agents and Chemotherapy*, 43, 4, (Apr 1999), pp. 846-849, ISSN 0305-7453.

Burnens, A., Stanley, J., Sechter, I., Nicolet, J. (1996). Evolutionary origin of a monophasic *Salmonella* serovar, 9,12:l,v:-, revealed by IS200 profiles and restriction fragment polymorphisms of the fljB gene. *Journal of Clinical Microbiology*, 34, 7, (Jul 1996), pp. 1641–1645, ISSN 0095-1137.

Carattoli, A., Tosini, F., Visca, P. (1998). Multidrug-resistant *Salmonella enterica* serotype Typhimurium infections. *The New England Journal of Medicine*. 339, (Sep 1998), pp. 921-922.

CDC. (2007a). Investigation of Outbreak of Human Infections Caused by *Salmonella* I 4,[5],12:i:-. In: *Centers for Disease Control and Prevention*. 15 July 2011. Available from: <http://www.cdc.gov/Salmonella/4512eyeminus.html>.

CDC (2007b). Turtle-Associated Salmonellosis in Humans - United States, 2006-2007. *Morbidity and Mortality Weekly Report*, 56, pp. 649-652.

CDC (2008). *Salmonella* Surveillance: Annual Summary, 2006. *US Department of Health and Human Services*. CDC, Atlanta, Georgia, USA.

de la Torre, E., Zapata, D., Tello, M., Mejia, W., Frias, N., Pena, F.J.G., Mateu, E.M., Torre, E. (2003). Several *Salmonella enterica* subsp *enterica* serotype 4,5,12:i: - Phage types isolated from swine originate from serotype typhimurium DT U302. *Journal of Clinical Microbiology*, 41, 6, (Jun 2003), pp. 2395-2400, ISSN 0095-1137.

del Cerro A., Soto S. M., Mendoza M. C. (2003). Virulence and antimicrobial-resistance gene profiles determined by PCR-based procedures for *Salmonella* isolated from samples of animal origin. *Food Microbiology*, 20, 24, (Aug 2003), pp. 431-438, ISSN 0740-0020.

Delgado R.N., Munoz Bellido J.L., García García M.I., Ibanez Perez R., Munoz Criado S., Serrano Heranz R., Saenz Gonzalez M.C., García Rodríguez J.A. (2006). Molecular epidemiology of drug-resistant *Salmonella* Typhimurium in Spain. *Revista Espanola de Quimioterapia.* 19, 2, (Jun 2006), pp. 152–160, ISSN 0214-3429.

Dionisi, A.M., Graziani, C., Lucarelli, C., Filetici, E., Villa, L., Owczarek, S., Caprioli, A., Luzzi, I. (2009). Molecular Characterization of Multidrug-Resistant Strains of *Salmonella enterica* Serotype Typhimurium and Monophasic Variant (S. 4,[5], 12:i:-) Isolated from Human Infections in Italy. *Foodborne Pathogens and Disease*, 6, 6, (Jul-Aug 2009), pp. 711-717, ISSN 1535-3141.

Echeita, M.A., Díez, R., Usera, M.A. (1999). Distribución de serotipos de *Salmonella spp.* aislados en España durante un periodo de 4 años (1993–1996), 1999. Enfermedades infecciosas y microbiologia clinica. 17, 1, pp. 9–14.

Echeita, M.A., Aladueña, A., Cruchaga, S., Usera, M.A. (1999). Emergence and Spread of an Atypical *Salmonella enterica* subsp. *enterica* Serotype 4,5,12:i:– Strain in Spain. *Journal of Clinical Microbiology.* 37, 10, (Oct 1999), pp. 3425.

Echeita, M.A., Herrera, S., Usera, M.A. (2001). Atypical, fljB-negative *Salmonella enterica* subsp *enterica* strain of serovar 4,5,12:i: appears to be a monophasic variant of serovar typhimurium. *Journal of Clinical Microbiology*, 39, 8, (Aug 2001), pp. 2981-2983, ISSN 0095-1137.

EFSA (2009). Analysis of the baseline survey on the prevalence of *Salmonella* in holdings with breeding pigs, in the EU, 2008, Part A: *Salmonella* prevalence estimates. *EFSA Journal*, 7, 12, (Dec 2008), pp. 1-93.

EFSA (2010a). Trends and Sources of Zoonoses and Zoonotic Agents and Food-borne Outbreaks in the European Union in 2008. *EFSA Journal*, 8, 1, (Jan 2010), pp. 1-48.

EFSA Panel on Biological Hazards (BIOHAZ). (2010b). Scientific Opinion on monitoring and assessment of the public health risk of "*Salmonella* Typhimurium-like" strains. *EFSA Journal.* 8, 10, (Oct 2010), pp. 1-48.

EFSA (2011). EU summary report on trends and sources of zoonoses and zoonotic agents and food-borne outbreaks 2009. *EFSA Journal.* 9, 3, (Mar 2011), pp.1-2.

EECDC; EFSA - Panel on Biological Hazards (BIOHAZ); CVMP-EMA; SCENIHR. (2009). Joint Opinion on antimicrobial resistance (AMR) focused on zoonotic infections. *EFSA Journal.* 7, 11, (Nov 2009), pp. 1-78.

Eswarappa, S.M., Janice, J., Nagarajan, A.G., Balasundaram, S.V., Karnam, G. (2008) Differentially Evolved Genes of *Salmonella* Pathogenicity Islands: Insights into the Mechanism of Host Specificity in *Salmonella*. *PLoS ONE*, 3, 12, (Dec 2008), pp. e3829, ISSN 1932-6203.

Ethelberg, S., Lisby, M., Torpdahl, M., Sorensen, G., Neimann, J., Rasmussen, P., Bang, S., Stamer, U., Hansson, H.B., Nygard, K., Baggesen, D.L., Nielsen, E.M., Molbak, K., Helms, M. (2004). Prolonged restaurant-associated outbreak of multidrug-resistant *Salmonella* Typhimurium among patients from several European countries. *Clinical Microbiology and Infection*, 10, 10, (Oct 2004), pp. 904–910, ISSN 1198-743X.

Foley, S.L., Lynne, A.M. (2008a). Food animal-associated *Salmonella* challenges: Pathogenicity and antimicrobial resistance. *Journal of Animal Science*, 86, 14, (Apr 2008), pp. E173–E187, ISSN 0021-8812.

Foley, S.L., Lynne, A.M., Nayak, R. (2008b). *Salmonella* challenges: Prevalence in swine and poultry and potential pathogenicity of such isolates. *Journal of Animal Science*, 86, 14, (Apr 2008), pp. E149-E162, ISSN 0021-8812.

Friedrich, A., Dorn, C., Schroeter, A., Szabo, I., Jaber, M., Berendonk, G., Brom, M., Ledwolorz, J., Helmuth, R. (2010). [Report on *Salmonella* isolates in livestock, food and feed, received at the German national reference laboratory for Salmonella during 2004-2008]. *Berliner und Münchener tierärztliche Wochenschrift*, 123, 7-8, (Jul-Aug 2010), pp. 265-277, ISSN 0005-9366.

Garaizar, J., Porwollik, S., Echeita, A., Rementeria, A., Herrera, S., Wong, R.M.Y., Frye, J., Usera, M.A., McClelland, M. (2002). DNA microarray-based typing of an atypical monophasic *Salmonella* enterica serovar. *Journal of Clinical Microbiology*, 40, 6, (Jun 2002), pp. 2074-2078, ISSN 0095-1137.

Garaizar, J., Porwollik, S., Echeita, A., Rementeria, A., Herrera, S., Wong, R. M.Y., Frye, J., Usera, M.A., McClelland, M M. (2003). DNA microarray-based typing of atypical monophasic *Salmonella* serovar (4,5,12:i:-) strains emergent in Spain. *Infection, Genetics and Evolution*, 2, pp. 286-287, ISSN 1567- 1348.

Gebreyes, W.A., Altier, C. (2002). Molecular characterization of multidrug-resistant *Salmonella enterica* subsp. *enterica* serovar Typhimurium isolates from swine. *Journal of Clinical Microbiology*, 40, 8, (Aug 2002), pp. 2813–2822, ISSN 0095-1137.

Grimont, P.A.D., Weill, F-X. (2007). *Antigenic formulae of the* Salmonella *serovars* (9th edition). WHO Collaborating Centre for Reference and Research on *Salmonella*. Institute Pasteur, Paris, France

Guerra, B., Laconcha, I., Soto, S.M., Gonzalez-Hevia, M.A., Mendoza, M.C. (2000). Molecular characterisation of emergent multiresistant *Salmonella enterica* serotype [4,5,12:i:-] organisms causing human salmonellosis. *FEMS Microbiology Letters*, 190, 2, (Sep 2000), pp. 341-347, ISSN 0378-1097.

Guerra, B., Soto, S.M., Arguelles, J.M., Mendoza, M.C. (2001). Multidrug resistance is mediated by large plasmids carrying a class 1 integron in the emergent *Salmonella* enterica serotype [4,5,12 : i :-]. *Antimicrobial Agents and Chemotherapy*, 45, 4, (Apr 2001), pp. 1305-1308, ISSN 0066-4804.'

Hampton, M.D., Threlfall, E.J., Frost, J.A., Ward, L.R., Rowe, B. (1995). *Salmonella* typhimurium DT 193: differentiation of an epidemic phage type by antibiogram, plasmid profile, plasmid fingerprint and *Salmonella* plasmid virulence (spv) gene probe. *Journal of Applied Bacteriology*, 78, 4 (Apr 1995), pp. 402-408, ISSN 0370-1778.

Hauser, E., Huhn, S., Junker, E., Jaber, M., Schroeter, A., Helmuth, R., Rabsch, W., Winterhoff, N., Malorny, B. (2009). Characterisation of a phenotypic monophasic variant belonging to *Salmonella enterica* subsp *enterica* serovar Typhimurium from wild birds and its possible transmission to cats and humans. *Berliner und Münchener tierärztliche Wochenschrift*, 122, 5-6, (May-Jun 2009), pp. 169-177, ISSN 0005-9366.

Hauser, E., Tietze, E., Helmuth, R., Junker, E., Blank, K., Prager, R., Rabsch, W., Appel, B., Fruth, A., Malorny. B. (2010a). Pork contaminated with *Salmonella enterica* serovar 4,[5],12:i:-, an emerging health risk for humans. *Applied and Environmental Microbiology*, 76, 14, (Jul 2010), pp. 4601-4610, ISSN 0099-2240.

Helaine, S., Thompson, J.A., Watson, K.G., Liu, M., Boyle, C., Holden, D.W. (2010). Dynamics of intracellular bacterial replication at the single cell level. *Proceedings of the National Academy of Sciences of the United States of America*, 107, (Feb 2010), pp. 3746- 3751, ISSN 0027-8424.

Hoelzer, K., Soyer, Y., Rodriguez-Rivera, L.D., Cummings, K.J., McDonough, P.L., Schoonmaker-Bopp, D.J., Root, T.P., Dumas, N.B., Warnick, L.D., Grohn, Y.T., Wiedmann, M., Baker, K.N., Besser, T.E., Hancock, D.D., Davis, M.A. (2010). The prevalence of multidrug resistance is higher among bovine than human *Salmonella enterica* serotype Newport, Typhimurium, and 4,5,12:i:- isolates in the United States but differs by serotype and geographic region. *Applied and Environmental Microbiology*, 76, 17, (Sep 2010), pp. 5947-5959, ISSN 0099-2240.

Hopkins, K.L., Kirchner, M., Guerra, B., Granier, S.A., Lucarelli, C., Porrero, M.C., Jakubczak, A., Threlfall, E.J., Mevius, D.J. (2010). Multiresistant *Salmonella enterica* serovar 4,[5],12:i:- in Europe: a new pandemic strain?. Eurosurveillance, 15, 22, (Jun 2010), pp. 1-9.

Hopkins, K.L., Nair, S., Kirchner, M., Guerra, B., Granier, S., Lucarelli, C., Porrero, C., Jakubczak, A., Threlfall, E.J., Mevius, D. (2010). Genetic variation in emerging multidrug-resistant *Salmonella enterica* 4,[5],12:i:- from seven European countries. *Proceedings of the 110th General Meeting of the American Society for Microbiology*, ISBN 1555816231, San Diego, CA, May 2010.

Ibarra, J.A., Steele-Mortimer, O. (2009). *Salmonella*--the ultimate insider. *Salmonella* virulence factors that modulate intracellular survival. *Cell Microbiology*, 11, 11, pp. 1579-1586, ISSN: 1462-5822

Kirk, M.D., Veitch, M.G., Hall, G.V. (2010). Gastroenteritis and food-borne disease in elderly people living in long-term care. *Clinical Infectious Diseases*, 50, 3 (Feb 2010), pp. 397-404, ISSN 1058-4838.

Laorden, L., Bikandi, J., Herrera-Leon, S., Sanchez, A., Echeita, A., Rementeria, A., Garaizar. J. (2009). Characterisation and evolutionary study of *Salmonella enterica* [4,5,12:i:-] using PFGE, MLVA, PCR and sequencing techniques. *Proceedings of the 3rd ASM Conference on Salmonella: Biology, Pathogenesis and Prevention*, ISBN 9781555815448, Aix-en-Provence, France, October 2009.

Laorden, L., Herrera-León, S., Sanchez, A., Bikandi, J., Rementeria, A., Echeita, A., Garaizar, J. (2010). Six types of deletions detected in monophasic *Salmonella enterica* 4,[5],12:i:- strains. *Proceedings of the 9th International Meeting on Molecular Epidemiological Markers*, Wernigerode, Germany, September 2010.

Levin, R.E. (2010). *Salmonella. Rapid Detection and Characterization of Foodborne Pathogens by Molecular techniques*, CRC Press, Taylor & Francis Group, ISBN 10 1420092421, Amherst, USA, pp. 79-138.

Lucarelli, C., Dionisi, A.M., Torpdahl, M., Villa, L., Graziani, C., Hopkins, K., Threlfall, J., Caprioli, A., Luzzi, I. (2010). Evidence for a second genomic island conferring multidrug resistance in a clonal group of strains of *Salmonella enterica* serovar Typhimurium and its monophasic variant circulating in Italy, Denmark, and the United Kingdom. *Journal of Clinical Microbiology*, 48, 6, (Jun 2010), pp. 2103-2109, ISSN: 0095-1137.

Machado, J., Bernardo, F. (1990). Prevalence of *Salmonella* in chicken carcasses in Portugal. *Journal of Applied Bacteriology*, 69, 4, (Oct 1990), pp. 477-480, ISSN 1364-5072.

Marcus, S.L., Brumell, J.H., Pfeifer, C.G., Finlay, B.B. (2000). *Salmonella* pathogenicity islands: big virulence in small packages. *Microbes and Infection*, 2, 2, (Feb 2000), pp. 145–156, ISSN 1286-4579.

Martins da Costa, P., Oliveira, M., Bica, A., Vaz-Pires, P., Bernardo, F. (2007). Antimicrobial resistance in *Enterococcus* spp. and *Escherichia coli* isolated from poultry feed and feed ingredients. *Veterinary Microbiology*, 120, 1-2, (Feb 2007), pp. 122–131, ISSN 0378-1135.

Mead, P.S., Slutsker, L., Dietz, V., McCaig, L.F., Bresee, J.S., Shapiro, C., Griffin, P.M., Tauxe, R.V. (1999). Food-Related Illness and Death in the United States. *Emerging Infectious Diseases*, 5, 5, (Oct 1999), pp. 607-625, ISSN 1080-6059.

Monroe, S., Polk, R. (2000). Antimicrobial use and bacterial resistance. *Current Opinion in Microbiology*, 3, 5, (Oct 2000), pp. 496–501, ISSN 1369- 5274.

Mossong, J., Marques, P., Ragimbeau, C., Huberty-Krau, P., Losch, S., Meyer, G., Moris, G., Strottner, C., Rabsch, W. Schneider, F. (2007). Outbreaks of monophasic *Salmonella enterica* serovar 4,[5],12:i:- in Luxembourg, 2006. *Eurosurveillance*, 12, 6, (Jun 2007), pp. 156-158, ISSN 1560-7917.

Pang, T., Levine, M.M., Ivanoff, B., Wain, J., Finlay, B.B. (1998). Typhoid fever—important issues still remain. *Trends in Microbiology*, 6, 4, (April 1998), pp. 131–133, ISSN 0966-842X.

Peters, T., Hopkins, K.L., Lane, C., Nair, S., Wain, J., de Pinna, E. (2010). Emergence and characterization of *Salmonella enterica* serovar Typhimurium phage type DT191a. *Journal of Clinical Microbiology*, 48, 9, (Sep 2010), pp. 3375- 3377, ISSN 0095-1137.

Popoff, M.Y., Le Minor, L. (1997). *Antigenic formulas of the Salmonella serovars* (7th edition). WHO Collaborating Centre for Reference and Research on *Salmonella*, Institut Pasteur, Paris, France.

Pornruangwong, S., Sriyapai, T., Pulsrikarn, C., Sawanpanyalert, P., Boonmar, S., Bangtrakulnonth, A. (2008). The epidemiological relationship between *Salmonella enterica* serovar typhimurium *and Salmonella enterica* serovar 4,[5],12:1:- isolates from humans and swine in Thailand. *Southeast Asian Journal of Tropical Medicine and Public Health*, 39, 2, pp. 288-296, ISSN 0125-1562.

Sala, A. (2002). Tracing the origin of monophasic serovar *S. enterica* I 4, 5,12: I:-: Genetic similarity with *S. enterica* typhimurium. *Medicina (Ribeirao Preto)*, 35, 1, 510-511, ISSN 2176-7262.

Saroj, S.D., Shashidhar, R., Karani, M., Bandekar, J.R. (2008). Distribution of *Salmonella* pathogenicity island (SPI)-8 and SPI-10 among different serotypes of *Salmonella*. Journal of Medical Microbiology, 57, 4, (Apr 2008), pp. 424-427, ISSN 0022-2615.

Sayah, R.S., Kaneene, J.B., Johnson, Y., Miller, R. (2005). Patterns of antimicrobial resistance observed in *Escherichia coli* isolates obtained from domestic- and wild animal fecal samples, human septage, and surface water. *Applied and Environmental Microbiology*, 71, 3, (Mar 2005), pp. 1394–1404, ISSN 0099-2240.

Soyer, Y., Switt, A.M., Davis, M.A., Maurer, J., McDonough, P.L., Schoonmaker-Bopp, D.J., Dumas, N.B., Root, T., Warnick, L.D., Grohn, Y.T., Wiedmann, M. (2009). *Salmonella enterica* serotype 4,5,12:i:-, an emerging *Salmonella* serotype that represents multiple distinct clones. *Journal of Clinical Microbiology*, 47, 11, (Nov 2009), pp. 3546-3556, ISSN 0095-1137.

Stevens, M.P., Humphrey, T.J., Maskell, D.J. (2009). Molecular insights into farm animal and zoonotic *Salmonella* infections. *Philosophical Transations of the Royal Society B*, 364, 1530, (Sep 2009), pp. 2709–2723, ISSN 0962-8436.

Switt, A.I.M., Soyer, Y., Warnick, L.D., Wiedmann, M. (2009). Emergence, distribution and molecular and phenotypic characteristics of *Salmonella enterica* serotype 4,5,12:i:-. *Foodborne Pathogens and Disease*, 6, 4, (May 2009), pp. 407–415, ISSN 1535-3141.

Tavechio, A.T., Fernandes, S.A., Ghilardi, A.C.R., Soule, G., Ahmed, R., Melles, C.E.A. (2009). Tracing lineage by phenotypic and genotypic markers in *Salmonella enterica* subsp *enterica* serovar I,4,[5],12:i:- and *Salmonella* Typhimurium isolated in state of Sao Paulo, Brazil. *Memórias Do Instituto Oswaldo Cruz*, 104, 7, (Nov 2009), pp. 1042-1046, ISSN 0074-0276.

Téllez, S., Briones, V., González, S., García-Pena, F.J., Altimira, A., Vela, A.I., Blanco, M.M., Ballesteros, C., Fernández-Garayzábal, J.F., Goyache, J. (2002). *Salmonella* septicaemia in a beauty snake (*Elaphe taeniura taeniura*): a case report. *The Veterinary Record*, 151, 1, (Jul 2002), pp. 28-29, ISSN 0042-4900.

Tierrez A., Garcia-del Portillo F. (2005). New concepts in *Salmonella* virulence: the importance of reducing the intracellular growth rate in the host. *Cellular Microbiology*, 7, 7, (Jul 2005), pp. 901-909, ISSN 1462-5822.

Torpdahl, M., Litrup, E., Nielsen, E.M. (2009). Caracterisation and prevalence of *Salmonella* 4,5,12:i:- in Denmark, the monophasic variant of *Salmonella* Typhimurium. *Proceedings of the 3rd ASM Conference on Salmonella: Biology, Pathogenesis and Prevention*, ISBN 9781555815448, Aix-en-Provence, France, October 2009.

Trupschuch, S., Gomez J.A.L., Ediberidze, I., Flieger, A., Rabsch, W. (2010). Characterisation of multidrug-resistant *Salmonella Typhimurium* 4,[5],12:i:- DT193 strains carrying a novel genomic island adjacent to the thrW tRNA locus. *International Journal of Medical Microbiology*, 300, 5, pp. 279-288, ISSN 1438-4221.

Van Pelt, W., Schimmer, B., Stenvers, O.F.J., Langelaar M.F.M. (2009). *Staat van zoönosen 2007-2008*. National Institute for Public Health and the Environment, pp. 64.

Vieira-Pinto, M., Temudo, P., Martins, C. (2005). Occurrence of *Salmonella* in the ileum, ileocolic lymph nodes, tonsils, mandibular lymph nodes and carcasses of pigs slaughtered for consumption. *Journal of Veterinary Medicine Serie B*, 52, 10, pp. 476-481, ISSN 0931-1793.

Yamamoto, S., Kutsukake, K. (2006). *fljA*-mediated posttranscriptional control of phase 1 flagellin expression in flagellar phase variation of *Salmonella enterica* serovar Typhimurium. *Journal of Bacteriology*, 188, 3, (Feb 2006), pp. 958–967, ISSN 0021-9193.

Zaidi, M.B., Leon, V., Canche, C., Perez, C., Zhao, S., Huber, S.K., Abbott, J., Blickenstaff, K., McDermott, P.F. (2007). Rapid and widespread dissemination of multidrug-resistant blaCMY-2 *Salmonella* Typhimurium in Mexico. *Journal of Antimicrobial Chemotherapy*, 60, (May 2007), pp. 398-401, ISSN 0305-7453.

Zamperini, K., Soni, V., Waltman, D., Sanchez, S., Theriault, E.C., Bray, J., Maurer, J.J. (2007). Molecular Characterization Reveals *Salmonella* enterica Serovar 4,[5],12:i:2-from Poultry Is a Variant Typhimurium Serovar. *Avian Diseases*, 51, 4, (Dec 2007), 958-964, ISSN 0005-2086.

Zhao, Y. (2001). *Virulence factors of Salmonella enterica serovar Enteritidis*. PhD Thesis. Faculty of Veterinary Medicine, Utrecht University, Utrech, The Netherlands, pp. 96.

2

The Different Strategies Used by *Salmonella* to Invade Host Cells

Rosselin Manon, Abed Nadia, Namdari Fatémeh,
Virlogeux-Payant Isabelle, Velge Philippe and Wiedemann Agnès
INRA Centre de Tours, UR1282 Infectiologie Animale et Santé Publique, Nouzilly
IFR 136, Agents Transmissibles et Infectiologie, Nouzilly
France

1. Introduction

Salmonella enterica are members of a Gram-negative enteropathogenic bacteria family, which are able to infect a great diversity of hosts, including human. According to serotypes and hosts, *Salmonella enterica* cause a wide range of food- and water-borne diseases ranging from self-limiting gastroenteritis to systemic typhoid fever. Moreover, no other known bacterial pathogens belonging to a single species show such a remarkable variation in their host specificity. Ubiquitous serotypes such as Typhimurium and Enteritidis tend to produce acute but self-limiting enteritis in a wide range of hosts, whereas host-specific serotypes are associated with severe systemic disease in healthy outbred adults of a single species that may not involve diarrhoea (e.g. Gallinarum in poultry). Host-restricted serotypes are primarily associated with systemic disease in one host (e.g. Dublin in cattle, Choleraesuis in pigs), but may cause disease in a limited number of other species (Velge *et al.*, 2005).

For all these serotypes, the intestinal barrier crossing constitutes a crucial step for infection establishment. As shown in Figure 1, *Salmonella* can induce their own entry into enterocytes, but M cells and CD18-expressing phagocytes also facilitate their translocation through the intestinal epithelium (Watson & Holden, 2010). During gastroenteritis pathology, host colonization is restricted to the intestinal tract. However, *Salmonella* also have the ability to disseminate to extra-intestinal sites at least via CD18-expressing phagocytes, leading to deep organ colonization (Vazquez-Torres *et al.*, 1999).

Bacterial pathogens have developed two different mechanisms to invade non-phagocytic host cells by hijacking physiological cellular processes. Bacteria, such as *Listeria monocytogenes* and *Yersinia pseudotuberculosis* express surface proteins that interact with receptor on the host cell plasma membrane. This interaction promotes an activation of host cell signaling pathways, leading to actin remodelling. This process is referred to as a Zipper mechanism and is characterized by the induction of little protrusive activity and thin membrane extensions (Figure 2A and C) (Cossart & Sansonetti, 2004). Other bacteria, such as *Shigella flexneri*, do not require a receptor but trigger internalization from "inside" via the action of pathogen-effector proteins delivered by specialized protein secretion systems (Schroeder & Hilbi, 2008). Translocated effector proteins effectively allow the bacteria to "hijack" many essential intracellular processes and induce a massive reorganization of the host actin cytoskeleton, resulting in intense membrane ruffling and internalization of the bacteria. This invasion process is referred to as a Trigger mechanism (Figure 2B and D) (Cossart & Sansonetti, 2004).

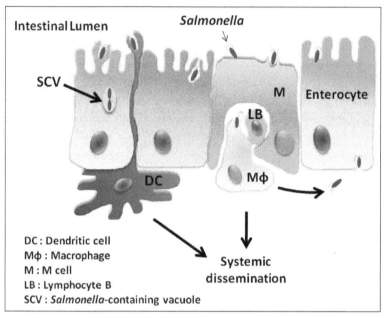

Fig. 1. Intestinal barrier crossing by *Salmonella enterica* through M cells, enterocytes or following a luminal capture by CD18+ phagocytes such as dendritic cells.

The reorganization of actin cytoskeleton at the entry site is a crucial step for Trigger and Zipper bacterial internalization. In eukaryotic cells, actin exists as a globular monomer (G-actin) which can assemble to form a filamentous structure (F-actin). In physiological conditions, actin polymerization requires different steps. First, nucleation of actin which consists in regrouping three actin monomers, is stimulated by cellular factors such as the Arp2/3 complex (Mullins *et al.*, 1998). Once nucleated, the addition of ATP-actin-monomers at the barbed extremity of the filaments allows actin elongation (Pollard *et al.*, 2000). The three-dimensional structure of actin filaments is ensured by capping proteins and other actin-binding proteins such as actinin, gelsolin, and villin that enable bundling of filaments (Bretscher, 1991; Hartwig & Kwiatkowski, 1991). Actin dynamics regulation is closely associated with small Rho guanosine triphosphatase protein (RhoGTPase) activity. RhoGTPases cycle between an inactive guanine di-phosphate (GDP)-bound form and an active guanine tri-phosphate (GTP)-bound form. The switch between inactive and active state is regulated by guanine exchange factors (GEF) which catalyze the exchange of GDP with GTP and GTPase activating proteins (GAP) which hydrolyze GTP into GDP to switch off their active state. When bound to GTP, Rho GTPases target and activate downstream effectors such as proteins from the Wiscott-Aldrich Syndrome protein (WASP) / N-WASP Family, leading to nucleator activation and actin reorganization. All these steps are required during bacterial internalization.

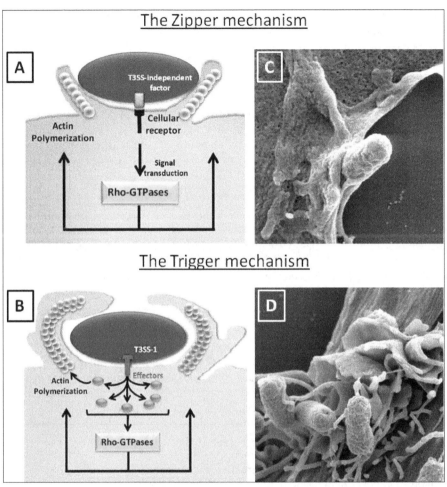

Fig. 2. Models of Zipper and Trigger invasion mechanisms. (A) The Zipper process is initiated by an interaction between a host cell receptor and a bacterial surface protein which allows the activation of RhoGTPases and actin polymerization at the entry site. (B) In contrast, during the Trigger mechanism, RhoGTPases are targeted by bacterial effectors which are directly translocated into host cell via a type–three secretion system, leading to actin polymerization and internalization. Electron scanning microscopy pictures show (C) *S.* Enteritidis invading fibroblasts via a Zipper process which is characterized by weak membrane rearrangements and (D) via a Trigger process which is characterized by intense membrane rearrangements.

The study of host cell invasion by *Salmonella* has been initiated in 1967 by Takeuchi (Takeuchi, 1967). For decades, it was described in the literature that *Salmonella* can enter cells only via a "Trigger" mechanism mediated by a type-three secretion system (T3SS-1) encoded by the *Salmonella* pathogenicity island-1 (SPI-1) (Ibarra & Steele-Mortimer, 2009). Recent data have showed that cell invasion could occur despite the absence of the T3SS-1

(Aiastui *et al.*, 2010; Radtke *et al.*, 2010; Rosselin *et al.*, 2011), indicating that the dominant paradigm postulating that a functional SPI-1/T3SS is absolutely required for cell entry, should be reconsidered. Moreover, the characterization of one T3SS-1-independent invasion pathway revealed that *Salmonella* have also the ability to enter cells via a Zipper process mediated by the Rck invasin (Rosselin *et al.*, 2011). Consequently *Salmonella* are the first bacteria found to be able to invade cells both via a Zipper and a Trigger mechanism.

Here, our current understanding of the different strategies used by *Salmonella* to invade host cells will be summarized and we will focus on how *Salmonella* are able to manipulate the host actin cytoskeleton, leading to discrete or intense membrane rearrangements. The gap of our knowledge about these different entry pathways will be discussed.

2. Invasion mechanism dependent on the T3SS-1

The Type-Three Secretion System (T3SS-1) is the best characterized invasion system of *Salmonella*. It allows bacterial internalization into non-phagocytic cells via a Trigger mechanism which induces massive actin rearrangements and intense membrane ruffling at the entry site (Cossart & Sansonetti, 2004). Under environmental conditions that enable the expression of the T3SS-1, the secretion apparatus is assembled at the bacterial surface and effectors are translocated into the eukaryotic cytosol following an interaction between the bacteria and the host cell (Garner *et al.*, 2002; Hayward *et al.*, 2005).

2.1 T3SS-1 structure

T3SSs are supramolecular complexes that play a major role in the virulence of many Gram-negative pathogens by injecting bacterial protein effectors directly into host cells in an energy-dependent (ATP) manner (Galan & Wolf-Watz, 2006). These complexes cross both inner and outer membranes of bacteria and are able to create a pore in eukaryotic membrane upon contact with a host cell. They are made of an exportation apparatus, a basal body, a needle and a translocon at the tip of the needle (Figure 3A). The structure of these T3SSs shows a high degree of conservation among pathogens (Tampakaki *et al.*, 2004) and the *Salmonella* T3SS-1 apparatus shares in particular a high homology with the T3SS of *Shigella*, also involved in host cell invasion (Groisman & Ochman, 1993).

The basal body of the T3SS anchors the complex into the bacterial inner and outer membranes (Figure 3A). It is composed of PrgH, PrgK and InvG proteins which assemble into an inner ring (PrgH and PrgK) and an outer ring (InvG) (Schraidt & Marlovits, 2011). Anchored to the basal body via its transmembrane part, the needle protrudes from the outer membrane as a long filament of 50 nm length and is composed of the single PrgI protein (Kimbrough & Miller, 2000). At the extremity of the needle, a complex of three proteins (SipB, SipC, SipD), known as the translocon, is able to form a pore in the eukaryotic target cell, allowing the secretion of effector proteins (Mattei *et al.*, 2011). SipB, SipC and SipD proteins (also referred to Ssp proteins) share homology with other translocon proteins such as IpaB, IpaC and IpaD proteins of *Shigella* (Hueck, 1998). SipD has a hydrophilic domain and interacts directly with the PrgI needle protein (Rathinavelan *et al.*, 2011) while the two other proteins of the translocon (SipB and SipC) have a hydrophobic domain and are therefore directly involved in the pore formation (Hayward *et al.*, 2000; Miki *et al.*, 2004). Particularly, it has been shown that the interaction of SipB with cellular cholesterol is necessary for effector translocation (Hayward *et al.*, 2005). Finally, the translocation of T3SS-1 effector proteins requires an exportation apparatus located at the inner membrane level and made of highly conserved proteins among T3SSs (SpaP, SpaQ, SpaR, SpaS, InvA, InvC and OrgB). The unfolded effectors in association with their chaperone are targeted to the exportation apparatus and the ATPase InvC produces the energy necessary to the transport of these proteins through the needle (Akeda & Galan, 2004). The appropriate hierarchy in the secretion process is established by a cytoplasmic sorting platform composed of SpaO, OrgA and OrgB (Lara-Tejero *et al.*, 2011). This platform sequentially loads the secreted proteins by interacting with their chaperones to ensure a specific order of secretion and optimize host cell invasion.

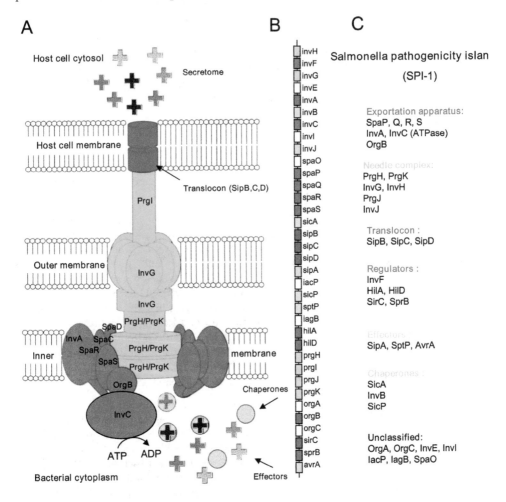

Adapted from Kimbrough and Miller, 2002.
A. Localization of the T3SS-1 structure proteins. **B.** Schematic representation of SPI-1 island encoding the T3SS-1 proteins. **C.** Functional classification of SPI-1-encoded proteins.

Fig. 3. Structure and organization of *Salmonella* T3SS-1.

2.2 Regulation of T3SS-1 expression

During *Salmonella* infection, a crucial step is the crossing of the intestinal barrier. The host environment encountered by the bacteria, and more particularly the small intestine environment, plays a major role in the invasion as it controls expression of the secretion apparatus. Coordination of T3SS-1 expression genes, almost all located on the *Salmonella* Pathogenicity Island 1 (SPI-1) is complex and well-timed. In response to different environmental stimuli, a sophisticated regulatory network controlling the expression of SPI-1 has been established (Figure 4). Our purpose here is not to set up the thorough state-of-the-art on all the regulators involved in *Salmonella* invasion, but to give a general overview of this system (for a review, see (Ellermeier & Slauch, 2007)).

SPI-1 contains 39 genes encoding structural T3SS-1 proteins (*inv/spa* and *prg* operon), translocon proteins (SipB, C, D), some effectors (SipA, SptP and AvrA), some chaperones (SicA, InvB, SicP) and finally four transcriptional regulators (HilA, hilC, HilD and InvF) (Figure 3B and C). Other genes encoding secreted effectors (*sopA, sopB, sopD, sopE, sopE2, slrP, sspH1, sspH2*) are located elsewhere on the chromosome.

HilA is central for SPI-1 transcriptional regulation. This protein activates directly the transcription of *prg*, *inv/spa* and *sip* operons, encoding structural components and some secreted effectors of T3SS-1 respectively. In addition, HilA induces the transcription of *invF*, encoding a transcriptional activator and targeting, among others, *sip* operon, *sopE* and *sopB* genes (Darwin & Miller, 1999). The sequential expression of HilA and InvF regulators allows a hierarchical regulation of invasion genes.

Then, a second crucial level of SPI-1 transcriptional regulation takes place through the regulation of HilA *via* a feed-forward loop, involving three homologous transcriptional regulators: HilC, HilD and RtsA. Each of them binds directly to the *hilA* promoter and is able to activate its own expression. In fact, HilC, HilD and probably RtsA, act as derepressors of *hilA* transcription by counteracting the silencing exerted by nucleoid-structuring proteins such as H-NS or Hha (Akbar *et al.*, 2003; Queiroz *et al.*, 2011). The reason why HilC, HilD and RtsA play such an important role in T3SS-1 expression through *hilA* regulation is that they are at the integration point of a lot of signals that control SPI-1 expression (Figure 4). In this regulatory circuit, it is currently admitted that HilD has a predominant role whereas HilC and RtsA simply act as signal amplifiers. However, it has also been shown that these three regulators are also directly implicated in the regulation of others invasion genes (Akbar *et al.*, 2003; Ellermeier & Slauch, 2004).

Moreover, besides these direct regulators, a great number of other *hilA* regulators, acting mainly through HilD, have been identified. Among them, two-component systems play a major role. They sense environmental conditions and allow the transmission of different signals which modulate T3SS-1 genes expression. Some are able to activate indirectly HilA expression such as BarA/SirA and OmpR/EnvZ, whereas others such as PhoP/PhoQ and PhoB/PhoR repress it. In fact, HilA expression can also be inhibited. HilE has been identified as a negative regulator of *hilA* transcription preventing HilD activity (Fahlen *et al.*, 2000; Baxter *et al.*, 2003). It has been suggested that the two-component systems PhoP/PhoQ and PhoB/PhoR act through HilE to regulate SPI-1 (Baxter & Jones, 2005).

As stated above, environmental signals play a major role in *Salmonella* invasion. Low oxygen tension, high osmolarity, high iron concentration, neutral pH are conditions found in the ileum, known to be the preferential invasion site of *Salmonella*. Thus, as expected in these conditions, invasion genes are activated through *hilA* expression. In contrast, when *Salmonella* are located at unfavorable sites for invasion in the host organism, the presence of signals such as bile, secreted into the proximal small intestine or cationic peptides, known to exist in macrophages, inhibits T3SS-1 expression (Figure 4).

Although much has already been identified about the regulation of SPI-1, recently, it became more evident that mechanisms regulating this system are more complex than previously thought.

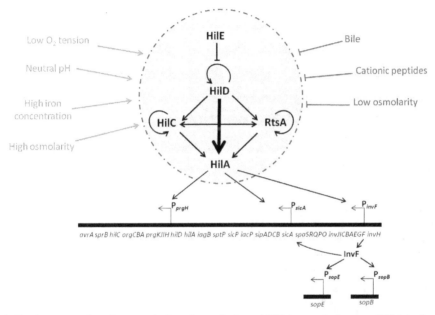

Fig. 4. Environmental and transcriptional regulation of SPI-1 encoded genes. HilA is the major regulator of SPI-1 and is itself regulated by other regulators such as HilC, HilD, RtsA and HilE. When *Salmonella* reach the small intestine, a low O_2 tension, a neutral pH, a high iron concentration and a high osmolarity activate SPI-1 expression. In contrast, the presence of bile or cationic peptides represses its expression.

2.3 Subversion of the cellular machinery during T3SS-1-dependent entry

Among the effectors that are translocated into host cell by the T3SS-1, six are essential to cell invasion (SipA, SipC, SopB, SopD, SopE, SopE2) while the other effectors contribute to a variety of post-invasion processes such as host cell survival and modulation of the inflammatory response (Patel & Galan, 2005). To trigger internalization into cells, effectors manipulate actin cytoskeleton either directly or indirectly. They also manipulate the delivery of vesicles to the site of bacterial entry to provide additional membrane and allow the extension and ruffling of the plasma membrane necessary to promote invasion. In later steps, membrane fission occurs to induce the sealing of the future *Salmonella*-containing vacuole (SCV) and actin filaments are depolymerized, enabling the host cell to recover its normal shape (Figure 5).

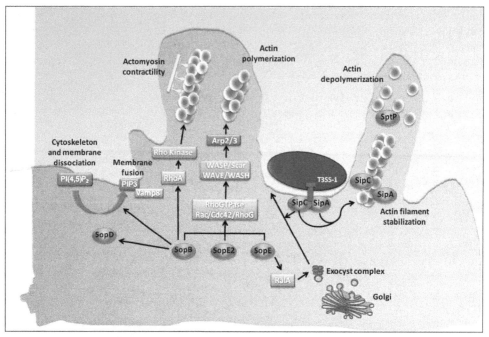

Fig. 5. Intracellular activities of the T3SS-1 effectors: SipA and SipC bind actin directly whereas SopE, SopE2 and SopB stimulate RhoGTPase activity. SipC and SopE also act in cooperation to recruit exocytic vesicles at the entry site. SopB plays diverse roles during invasion: it promotes actomyosin contractility and changes phosphoinositide concentrations to facilitate the dissociation between actin cytoskeleton and membrane at the entry site. SopB also probably triggers the delivery of vesicles to the bacterial entry site in a VAMP-8 dependent way and activate SopD which seems to contribute to the sealing and the formation of the SCV. The last effector SptP turns off the activity of RhoGTPases to allow the host cell to regain its initial shape.

2.3.1 Actin manipulation

Actin polymerization is an essential process induced by T3SS-1 effectors during entry. Some effectors stimulate actin rearrangements indirectly and two effectors, SipA and SipC, manipulate actin directly. SipA and SipC are localized at the eukaryotic plasma membrane during entry (Scherer et al., 2000) but they are also translocated into the host cell cytoplasm (Hueck et al., 1995; Kaniga et al., 1995).

SipC, as part of the translocon, is required for T3SS-1 effector translocation and involved in actin nucleation and bundling. Due to its various functions, a 95% decrease in invasion is observed when *sipC* is deleted (Chang et al., 2005). The C-terminal region of SipC (amino-acids A321-A409) encodes the effector translocation activity (Chang et al., 2005) and its central region (amino-acids N201–S220) binds to actin and induces a rapid nucleation and elongation of actin filaments. More precisely the regions containing the amino acids from H221–M260 and L381–A409 bind to and bundle actin filaments *in vitro* (Myeni & Zhou,

2010). The bundling activity of SipC is essential for internalization as a *sipC* mutant lacking bundling activity is impaired in cell invasion. SipC dimerization / multimerization seems to be required for nucleation and a *sipC* mutant deficient for multimerization and actin nucleation failed to cause severe colitis in a mouse model (Chang *et al.*, 2007).

As SipC, SipA binds directly to actin (amino acids P446-R685) and modulates cytoskeleton dynamics in different ways (Galkin *et al.*, 2002; Higashide *et al.*, 2002). A *Salmonella* Typhimurium *sipA* mutant exhibits a 60-80% decrease in invasion compared to the wild-type strain (Perrett & Jepson, 2009), which can be correlated to the fact that SipA is a multifunctional effector. First, Zhou *et al.* (1999b) have observed that SipA reduces the critical concentration of G-actin in the cytosol required for polymerization. In addition, to facilitate polymerization, SipA also stabilizes actin filaments by displacing ADF / cofilin factor which stimulates actin depolymerization and by protecting F-actin from gelsolin severing (McGhie *et al.*, 2004). Moreover, SipA enhances actin cross-linking both by interacting with T-Plastin, a cellular bundling protein, and by enhancing SipC activity (Zhou *et al.*, 1999a; McGhie *et al.*, 2001). Interestingly, Perrett and Jepson have demonstrated that a *sipA* deletion induces a decrease in host cell invasion but without affecting the frequency of membrane ruffle formation (Perrett & Jepson, 2009). By visualizing the membrane ruffles in the absence of SipA, they have observed that SipA is required to ensure the spatial localization of actin rearrangement beneath invading *Salmonella* for efficient uptake of bacteria.

In contrast to SipA and SipC, SopB, SopE and SopE2 exert their activity into host cells by inducing actin polymerization in an indirect way. They activate RhoGTPases which are key cellular effectors that regulate actin cytoskeleton remodeling (Patel & Galan, 2005).

During *Salmonella* T3SS-1 dependent invasion, SopE and SopE2 effectors mimic GEF activity to activate RhoGTPases. Their activation induces actin polymerization by stimulating downstream cellular proteins such as N-WASP, WAVE and WASH which activate the Arp2/3 nucleator complex (Buchwald *et al.*, 2002; Schlumberger *et al.*, 2003). SopE specifically targets the Rho GTPases Rac and Cdc42 *in vitro* whereas SopE2 only activates Cdc42 (Friebel *et al.*, 2001). The role of Rac in *Salmonella* T3SS-1-dependent entry is well characterized but the role of Cdc42 is controversial. Criss *et al.* (2001) have demonstrated by using dominant negative forms and by pull-down assays that, in contrast to non-polarized cells, Cdc42 is not required and not activated during invasion of MCDK polarized epithelial cells. In the same way, Patel and Galan have observed that the depletion of Cdc42 by RNA interference (RNAi) in COS-2 and Henle-407 had no effect on membrane ruffling and efficient uptake (Patel & Galan, 2006). In contrast, a recent genome-scale RNAi screening in HeLA cells indicated that Cdc42 depletion induced a 70% decrease in invasion, suggesting that it is required for the T3SS-1-dependent invasion process (Misselwitz *et al.*, 2011). All together, it is difficult to conclude about the involvement of Cdc42 during T3SS-1-dependent entry.

Salmonella mutant strains lacking both *sopE* and *sopE2* are still able to invade cells in a SopB dependent way (Zhou *et al.*, 2001). SopB is an inositol phosphatase which shares homology with eukaryotic phopho-inositol (PI) phosphatases (Norris *et al.*, 1998). Like SopE and SopE2, SopB targets a GTPase from the Rho family, RhoG, but in an indirect manner since SopB activates the GEF protein that allows RhoG activation (Patel & Galan, 2006). Once activated, RhoG induces actin polymerization at the entry site presumably by stimulating the Arp2/3 complex (Patel & Galan, 2006). A *sopB* deletion in *S.* Typhimurium induces a 50% decrease in invasion. But when *sopB* deletion is coupled with a *sopE* deletion, *Salmonella* uptake is drastically impaired (Zhou *et al.*, 2001), demonstrating that all the T3SS-1 effectors work in concert to trigger entry.

Recently, SopB was also shown to manipulate actomyosin contractility to mediate invasion. SopB recruits myosin II by activating RhoA and its Rho kinase downstream effector. In contrast to the process leading to actin polymerization during *Salmonella* entry, myosin II recruitment at the entry site is independent on Arp2/3 nucleator activity (Hanisch *et al.*, 2011).

In addition, different cellular proteins are involved during the T3SS-1 dependent entry of *Salmonella* without the identification of the bacterial effector responsible of this effect. This includes the focal adhesion kinase FAK, the Abelson tyrosine kinase Abl, and Shank3 (Shi & Casanova, 2006; Huett *et al.*, 2009; Ly & Casanova, 2009). During *Salmonella* uptake, FAK acts as a scaffolding protein, but not as a protein tyrosine kinase and its interaction with p130Cas is involved in actin reorganization and membrane ruffle formation (Shi & Casanova, 2006). But how *Salmonella* can nucleate assembly of focal adhesion-like complexes is still unclear and further research is needed to determine if this mechanism involves secreted bacterial effector proteins, other transmembrane host proteins, or both. Another scaffolding protein Shank3 also seems to regulate actin rearrangement during entry but the mechanisms leading to its recruitment and its activation have to be elucidated (Huett *et al.*, 2009). As the experiments were performed using HeLa cells which are highly permissive to the T3SS-1-dependent entry and with bacterial culture conditions that allow T3SS-1 expression, it is more probable that the *Salmonella* entry process inducing Shank3 recruitment is dependent on the T3SS-1. However, a T3SS-1-independent invasion process could also be involved because no mutant deficient for T3SS-1 expression was used as a control to analyze the involvement of Shank3.

Abelson tyrosine kinase Abl is also involved during *Salmonella* invasion as well as its effectors Abi1, a member of the WAVE2 complex, and CrkII (Ly & Casanova, 2009). Abi1 could thus enhance actin polymerization at the entry site but the role of CrkII during the invasion process remains poorly characterized.

2.3.2 Subversion of exocytosis machinery and membrane fusion during entry

Recent data indicate that membrane fusion is a major process involved during entry, suggesting that membrane ruffling requires the addition of intracellular membrane. A study focusing on the subversion of the host exocyst complex during *Salmonella* entry has showed cooperation between SipC and SopE (Nichols & Casanova, 2010). The exocyst is an octomeric complex (Sec3, Sec5, Sec6, Sec8, Sec10, Sec15, Exo70, Exo84) involved in vesicular trafficking which directs post-Golgi vesicules at specific site on the plasma membrane prior to their fusion. Nichols and Casanova have demonstrated that the mature exocyst complex is recruited at the entry site through an interaction between its subunit Exo70 and SipC. A depletion of exo70 or Sec5, another component of the exocyst complex, impairs *S.* Typhimurium invasion. Moreover, they have shown that SopE activates the Ras-related protein RalA, a small GTPase required for exocyst complex assembly (Nichols

& Casanova, 2010). It appears that SopE and SipC effectors manipulate the host exocyst to bring new membrane at the entry site in order to allow the formation of membrane ruffles and internalization.

SopB is also involved in this membrane fusion process through its inositol phosphatase activity. Dai *et al.* (2007) have shown that SopB-generated PI3P at the entry site leads to the recruitment of VAMP8, a host V-SNARE protein that mediates fusion between early and late endosomes. Moreover, depletion of VAMP8 by RNAi induces a decrease in invasion rate of a wild-type *S.* Typhimurium strain which is equivalent to that obtained with a *sopB* mutant. Thus, it seems that SopB promotes invasion by manipulating eukaryotic vesicular trafficking probably to induce fusion of intracellular vesicles to the cell membrane at the entry site.

How these events of vesicle-membrane fusion interact with actin cytoskeleton rearrangements to trigger entry has not been investigated yet. These different processes may synergize to induce internalization since actin dynamics is closely related to the metabolism of phosphoinosides (Honda *et al.*, 1999; Sechi & Wehland, 2000). However, VAMP8 which seems to be a marker of membrane fusion involved during *S.* Typhimurium invasion does not co-localize with F-actin during entry (Dai *et al.*, 2007). In addition, it could be interesting to better characterize the role of the cellular factor IQGAP1 which is required for *Salmonella* uptake and acts following an interaction with actin and the RhoGTPases Rac1 and Cdc42 (Brown *et al.*, 2007). Indeed, IQGAP1 is known to regulate actin architecture and interestingly, it also seems to act as a regulator of exocytosis by interacting with Exo70 (Rittmeyer *et al.*, 2008). IQGAP1 could thus be one of the missing links between actin rearrangement and membrane fusion during *Salmonella* entry. Further studies could overcome this issue.

2.3.3 Phagosome closure and restoration of the host cell normal shape

As described above, inositol phosphatase activity of SopB drives to changes in cellular phosphoinosite concentrations at the bacteria/cell contact. In addition to generate PI3P at the entry site, translocation of SopB into host cells also induces $PI(4,5)P_2$ hydrolysis, which leads to an almost complete absence of $PI(4,5)P_2$ at the membrane invagination regions (Terebiznik *et al.*, 2002). By reducing the local concentration of $PI(4,5)P_2$, SopB destabilizes cytoskeleton-plasma membrane interactions, thus reducing the rigidity of the membrane and promoting invasion by facilitating the fission and the sealing of the future *Salmonella*-containing vacuole. In addition to SopB, SopD also contributes to membrane fission. Boonyom *et al.* have demonstrated that a *sopD* deletion, like the *sopB* mutant, leads to a decrease in membrane fission during invasion and that SopD is recruited at the bacterial invasion site dependently on the phosphatase activity of SopB (Bakowski *et al.*, 2007). Thus, SopD seems to cooperate with SopB and contribute to *Salmonella* uptake by facilitating membrane fission at the entry site leading to the formation of the SCV (Bakowski *et al.*, 2007).

Following the formation of intense membrane ruffling and internalization, the eukaryotic cell regains its normal shape, inducing the closure of the vacuole of endocytosis containing the bacteria. The restoration of actin cytoskeleton is promoted by the effector SptP, a tyrosine phosphatase which inactivates the RhoGTPases Rac-1 and Cdc42 by stimulating their intrinsic GTPase activity (Fu & Galan, 1998; Fu & Galan, 1999). The N- terminal region of SptP interacts with Rac-1 and Cdc42 and a structural study of SptP indicates that this effector mimics the activity of host cell GAPs factors (Stebbins & Galan, 2000). Interestingly, SptP is regulated by two different mechanisms in order to delay its activity in host cell compared to that of SopE or SipA. A microscopy analysis revealed that SipA is injected earlier than SptP in the host cytoplasm, which would imply that SipA has a higher affinity for the exportation apparatus of the T3SS-1 than SptP (Winnen *et al.*, 2008). Moreover, SptP degradation by the host cell proteasome occurs later than SopE degradation (Kubori & Galan, 2003).

2.4 T3SS-1 contribution to *Salmonella* pathogenesis

The T3SS-1 is the main invasion factor of *Salmonella in vitro*. Nevertheless, its contribution to pathogenesis depends on the model used. *In vivo* studies with S. Dublin and S. Typhimurium serotypes have demonstrated that the T3SS-1 is essential for intestinal colonization and is required to induce enterocolitis in bovine, rabbit and murine models (Wallis & Galyov, 2000). In contrast, recent studies demonstrate that different serotypes of *Salmonella* lacking T3SS-1 still have the ability to invade *in vitro* cells of diverse origins and can be pathogenic in different *in vivo* infection models (Aiastui *et al.*, 2010; Rosselin *et al.*, 2011). In addition, it was shown that the T3SS-1 is not required for *Salmonella* internalization into a 3-Dimensional intestinal epithelium (Radtke *et al.*, 2010). Moreover, a SPI-1 mutant of S. Gallinarum exhibits a reduced invasiveness into avian cells but is fully virulent in adult chicken (Jones *et al.*, 2001). In S. Enteritidis and S. Typhimurium, the T3SS-1 is not essential during systemic infection of one week-old chicken or BalB/c mouse nor during the intestinal colonization of rabbit ileal loops (Coombes *et al.*, 2005; Jones *et al.*, 2007; Karasova *et al.*, 2010). Moreover, S. Senftenberg strains lacking SPI-1 are isolated from human clinical cases, suggesting that the T3SS-1 is dispensable by this serotype for the establishment of infection in humans (Hu *et al.*, 2008).

Taken together, these results indicate that T3SS-1- independent invasion mechanisms also play an important role in *Salmonella* infection and pathogenesis.

3. Invasion mechanisms independent of the T3SS-1

A *Salmonella* mutant, unable to express its T3SS-1 is still able to invade numerous cell lines and cell types and is shown to induce both intense and local membrane rearrangements (Rosselin et al., 2011). However, to date, little is known about the entry factors mediating these T3SS-1 independent invasion mechanisms. Here, we describe and sum up the state-of-art regarding these new invasion systems. Rck, PagN and HlyE are the three invasins identified as involved in *Salmonella* uptake. Moreover, Rosselin *et al.* (2011) have described that others unknown invasion factors exist although they are still not identified.

3.1 The Rck invasin

Among invasins that play a role in *Salmonella* invasion in a T3SS-1-independent way, Rck is clearly the best characterized. Rck is an 17kDa outer membrane protein (OMP) encoded by the large virulence plasmid of S. Enteritidis and S. Typhimurium (Heffernan *et al.*, 1992; Rotger & Casadesus, 1999). In addition to its ability to induce adhesion to and invasion of eukaryotic cells, Rck confers a high resistance level to complement killing by preventing the formation of the membrane attack complex (Heffernan *et al.*, 1992; Cirillo *et al.*, 1996; Rosselin *et al.*, 2010).

3.1.1 Rck structure

Rck is a member of an outer membrane protein family named "Ail/Lom family". This family consists of five members (Rck, Ail, Lom, OmpX and PagC) which are predicted to have eight transmembrane beta-sheets and four cell surface-exposed loops. Even if these proteins present a similar conformation, they have different functions. Rck and Ail (encoded by a chromosomal gene of *Yersinia Enterocolitica*) share the ability to promote serum resistance and epithelial cell invasion. These proteins do not exhibit homologous regions that could be related to these two identical roles. In Ail, the cell invasion property is associated with loop2 whereas loop3 and more precisely the region from the amino acids G113 to V159 was shown to be the minimal region of Rck required and sufficient for cell adhesion and invasion (Miller *et al.*, 2001; Rosselin *et al.*, 2010).

Another member of this family involved in virulence is PagC which is encoded by a *phoP*-regulated gene on the *Salmonella* chromosome and plays a role in intracellular macrophage survival in *Salmonella* (Miller *et al.*, 1989; Gunn *et al.*, 1995). Others members of this family are OmpX of *Enterobacter cloacae* (Mecsas *et al.*, 1995) and Lom, a protein expressed by bacteriophage λ from lysogenic *E. coli* (Reeve & Shaw, 1979) but none of them have known virulence-associated phenotype.

3.1.2 Rck regulation

A genetic screening performed in *S.* Typhimurium to identify genes regulated by SdiA (suppressor of division inhibition), a transcriptional regulator of quorum sensing, has suggested that *rck* belongs to a putative operon called the "*rck* operon" whose expression is activated by SdiA in an Acyl Homoserine Lactone (AHLs)-dependent manner (Figure 6) (Ahmer *et al.*, 1998; Michael *et al.*, 2001).

The *rck* operon contains 6 open reading frames: *pefI, srgD, srgA, srgB, rck* and *srgC* (Figure 6). Two genes in this operon, *pefI* (plasmid encoded fimbriae) and *srgA* (sdiA-regulated gene), affect the expression and function of the *pef* operon located upstream of the *rck* operon and involved in the biosynthesis of the Pef fimbriae. *pefI* encodes a transcriptional regulator of the *pef* operon, and SrgA is a DsbA paralog that efficiently oxidizes the disulfide bond of PefA, the major structural subunit of the Pef fimbriae (Bouwman *et al.*, 2003). These fimbriae are involved in biofilm formation, adhesion to murine small intestine and fluid accumulation in the infant mouse (Baumler *et al.*, 1996; Ledeboer *et al.*, 2006). Also localized on the *rck* operon, *srgD* encodes a putative transcriptional regulator. Recently, it has been shown that SrgD acts in cooperation with PefI to induce a synergistic negative regulation of flagellar genes expression (Wozniak *et al.*, 2009; Wallar *et al.*, 2011). The remaining genes on the *rck* operon have unknown functions and encode a putative outer membrane protein, SrgB, and a putative transcriptional regulator, SrgC.

Another locus regulated by SdiA-AHLs has been identified during screening. This chromosomal locus encodes a single gene named *srgE* (STM1554) (Ahmer *et al.*, 1998). No function for SrgE is described but a computational approach has suggested that SrgE is secreted by the T3SS-1 (Samudrala *et al.*, 2009) (Figure 6).

As *E. coli* and *Klebsiella*, *Salmonella* lack an AHL synthase and thus do not produce AHLs. However, SdiA can detect and bind AHLs produced by others bacterial species (Michael *et al.*, 2001). SdiA is a LuxR homologue and has two functional domains. The C-terminal region contains a predicted helix-turn-helix motif implicated in DNA binding and a N-terminal domain called "autoinducer domain" that interacts with AHLs. By NMR analysis, Yao *et al.* (2007) have shown that a direct interaction between SdiA and AHLs is required for SdiA folding and function.

Temperature also affects *rck* operon expression. At temperature below 30°C, the transcription of *rck* operon is repressed, while *srgE* is repressed only at temperature below 22°C (Smith & Ahmer, 2003)(Figure 6). As SdiA expression is not temperature regulated, another level of *rck* operon regulation remains to be identified.

In addition, a transcriptomic study has shown that Hha and its paralogue YdgH could be involved in the regulation of the *rck* operon (Vivero *et al.*, 2008).

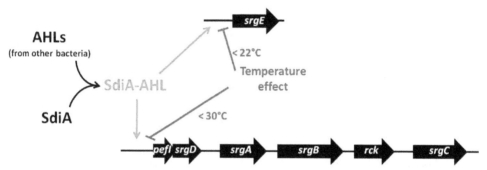

Fig. 6. Regulation of *rck* operon expression. When bound to AHLs, SdiA activates the expression of *rck* operon and *srgE*. Under temperatures that are lower than 30°C or 22°C, the expression of *rck* operon and *srgE* is inhibited, respectively.

3.1.3 Rck-mediated entry mechanism

When a *rck* mutant is grown under swarming conditions known to induce SdiA expression, a 2-3 fold decrease in epithelial cell invasion has been observed compared to the wild-type strain (Rosselin *et al.*, 2010). Moreover, in standard culture conditions, Rck overexpression leads to an increase in invasion.

By using both an initially non-invasive *E. coli* strain overexpressing Rck and latex beads coated with the minimal region of Rck inducing invasion (G113-V159), it was demonstrated that Rck alone is able to induce entry by a receptor-mediated process. This mechanism promotes local actin remodelling and weak and closely adherent membrane extensions (Rosselin *et al.*, 2010). *Salmonella* can thus enter cells through two distinct mechanisms: the Trigger mechanism mediated by its T3SS-1 apparatus and a Zipper mechanism induced by Rck. A model of this Zipper entry process is shown in figure 7. Following an interaction between Rck and its unknown cellular receptor, it was shown by using specific drugs and a dominant negative form that the class I PI3-kinase made of the p85-p110 heterodimer is required for Rck mediated entry. Moreover, Rck induces an increase in the interactions between p85 and phosphotyrosine residues, leading to the class I PI3-Kinase activation. Pharmacological approaches or Akt knockout cells also demonstrate that Akt is necessary to Rck-mediated internalization. Probably by binding to PI(3,4,5)P$_3$, Akt is recruited at the entry site and activated in a PI3-Kinase dependent way (Mijouin *et al.*, 2012).

The GTPase Rho is not involved during the Rck entry process but the use of dominant negatives demonstrates that Rac1 and Cdc42 are required (Rosselin *et al.*, 2010). Moreover, Rac1 is recruited at the entry site and Rck induces an increase in the level of active Rac1, demonstrating that it activates this GTPase (Rosselin *et al.*, 2010; Mijouin *et al.*, 2012). Rac1 activation occurs downstream on the PI-3kinase activity. Finally, overexpression of inhibitory constructs has shown that actin polymerization is dependent on the Arp2/3 nucleator complex during Rck-mediated entry (Rosselin *et al.*, 2010).

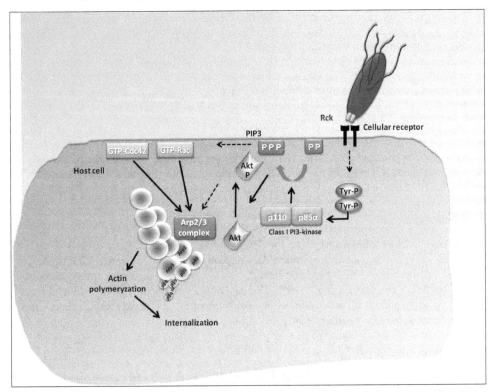

Fig. 7. Model of the cellular signaling pathway induces by Rck and leading to invasion. This Zipper entry process involves different cellular proteins: Class I PI3-Kinase, Akt, Rac and Cdc42, the Arp2/3 complex and actin. Dotted arrows: possible signalling events and/or interactions.

3.1.4 Rck contribution to *Salmonella* pathogenesis

The role of Rck in *Salmonella* invasion is clearly demonstrated *in vitro*, but its role in *Salmonella* pathogenesis is poorly understood. Indeed, *in vivo* conditions allowing *rck* expression are unclear. The fact that Rck is regulated by SdiA, a quorum sensing regulator suggests an intestinal role of this invasin (Ahmer *et al.*, 1998). However, the mechanisms leading to SdiA activation and *rck* expression are not well characterized. SdiA activation has been investigated in different hosts including rabbit, guinea pig, cow, turtle, mouse, pig and chicken but no activated-SdiA has been detected in the gastro-intestinal tract of these animals, except for turtles which were found to be co-infected with AHLs-producing *Aeromonas hydrophila* (Smith *et al.*, 2008). Another work has demonstrated that SdiA is activated in mice previously infected with *Yersinia enterocolitica*, which is able to synthesize AHLs (Dyszel *et al.*, 2010). However, in these conditions, SdiA activation does not confer a fitness advantage for *Salmonella* intestinal colonization in comparison to a *sdiA* mutant (Dyszel *et al.*, 2010). These results suggest that even if SdiA activation is achieved when AHL-producing strains colonize the gastrointestinal tract, it is not always sufficient to induce the expression of its regulon including *rck*. To assess the role of SdiA and its regulon during intestinal infection, Dyszel *et al.* (2010) have constructed a *Salmonella* strain able to produce AHLs. After co-infection of mice with the AHL-producing *Salmonella* strain and a *sdiA* mutant, it was shown that the constant activation of SdiA confers a selective advantage to *Salmonella*. Moreover, a loss of this selective advantage was observed when all individual SdiA-regulated genes were deleted, including *rck*, suggesting a role during intestinal colonization. Nevertheless, an *in vivo* model allowing a physiological activation of SdiA would be needed to assess the contribution of Rck to intestinal infection.

In addition, as *rck* is also regulated by an unidentified SdiA-independent system (Smith *et al.*, 2008), it is conceivable that Rck invasion mechanism is not restricted to the gastrointestinal tract. Considering that Rck is also involved in the resistance to complement killing, a systemic function of Rck is possible.

3.2 The PagN invasin

In addition to the T3SS-1 and Rck, the PagN outer membrane protein has also been identified as being involved in *Salmonella* invasion (Lambert & Smith, 2008). *pagN*, is localized on the centisome 7 genomic island and is widely distributed among *Salmonella enterica* serotypes (Folkesson *et al.*, 1999). The *pagN* ORF was originally identified during a Tn*phoA* random-insertion screening in *S.* Typhimurium performed to discover PhoP-activated genes (Belden & Miller, 1994).

3.2.1 PagN structure

PagN shares similarity to both the Tia and Hek invasins of *E. coli*, and presents 39% and 42% identity in amino acids with these two invasins, respectively. Tia and Hek are predicted to have eight transmembrane regions, four long exposed extracellular loops and three short periplasmic turns (Mammarappallil & Elsinghorst, 2000; Fagan *et al.*, 2008). Thus, PagN probably adopts a similar conformation as that of Hek and Tia.

3.2.2 PagN regulation

The *pagN* (*phoP*-activated gene) gene is *phoP*-regulated. The PhoP/PhoQ system is a two-component transcriptional regulatory system which modulates transcription of a multitude of virulence genes in *Salmonella*. This regulatory system is composed of the PhoQ sensor kinase (located at the membrane) and the PhoP response regulator. In response to specific stimuli such as acidified macrophage phagosome environment or low Mg^{2+} concentration, PhoQ is auto-phosphorylated and transfers its phosphates to the cytoplasmic DNA-binding protein PhoP, regulating specific genes.

3.2.3 PagN-mediated entry mechanism

Lambert & Smith (2008) have demonstrated that *pagN* deletion in *S.* Typhimurium leads to a 3-fold decrease in invasion of enterocytes without altering cell adhesion. At the cellular level, the PagN-mediated entry process is poorly characterized. It was only shown that actin polymerization is required during invasion (Lambert & Smith, 2008) and that PagN is able to interact with extracellular heparin proteoglycans (Lambert & Smith, 2009). However, proteoglycans cannot transduce a signaling cascade so it is probable that they act as co-receptors for invasion and not as the receptor itself. Moreover, although an interaction between PagN and heparin has been suggested, no clear heparin-binding motif was detected. Moreover, all PagN loops are required for invasion in epithelial cells (Lambert & Smith, 2009).

3.2.4 PagN contribution to *Salmonella* pathogenesis

PagN is required for survival in BALB/c mice (Heithoff *et al.*, 1999) and a *pagN* mutant is less competitive to colonize the spleen of mice compared to its parental strain (Conner *et al.*, 1998). However, the role of PagN in *Salmonella* pathogenesis is still unclear. *pagN* is activated by PhoP and thus maximally expressed intracellularly, a condition in which the SPI-1 island encoding the T3SS-1 is downregulated (Conner *et al.*, 1998; Heithoff *et al.*, 1999; Eriksson *et al.*, 2003). Lambert & Smith (2008) thus postulate that bacteria which exit epithelial cells or macrophages have an optimal level of PagN expression, but have a low T3SS-1 expression and this might facilitate subsequent interactions with mammalian cells that the pathogen encounters after host cell destruction.

3.3 The HlyE invasin

The *hlyE* gene is localized on the *Salmonella* pathogenicity island SPI-18 and is expressed by serotypes associated with systemic infection in humans including *S.* Typhi and *S.* Paratyphi A (Fuentes *et al.*, 2008). The *hlyE* sequence shares more than 90% identity with that of *Escherichia coli* HlyE (ClyA) hemolysin.The HlyE protein is able to lyse epithelial cells when exported from bacterial cell via outer membrane vesicule release (Wai *et al.*, 2003). Recently, Fuentes *et al.* (2008) have demonstratred that HlyE contributes to epithelial cell invasion of *S.* Typhi. However, the cellular events leading to HlyE-mediated invasion have not been characterized.

In vivo studies have shown that HlyE contributes to colonization of mouse deep organs (Fuentes *et al.*, 2008). However, how HlyE participates in establishment of systemic infection of *Salmonella* is not well understood.

3.4 Non-identified invasion factors

Several studies have revealed that invasion systems in *S.* Typhimurium and *S.* Enteritidis are not restricted to the T3SS-1, Rck and PagN. Indeed, a strain which does not express *rck*, *pagN* and the T3SS-1 is still able to significantly invade fibroblasts, epithelial and endothelial cells (Rosselin *et al.*, 2011). Results obtained by Aiastui *et al.* (2010) and van Sorge *et al.* (2011) have reinforced the idea that non-identified invasion factors are involved during entry into these cell types since *Salmonella* strains that cannot express the

T3SS-1 still enter into epithelial cells, endothelial cells and fibroblasts in a significant way. Moreover, invasion analyses of a 3-D intestinal epithelium by *S.* Typhimurium have also highlighted the fact that *Salmonella* possess non-characterized invasion factors (Radtke *et al.*, 2010).

Actin network remodeling and membrane rearrangements induced by these unknown factors have been visualized by confocal and electron microscopy as well as both Zipper-like and Trigger-like membrane alterations (Rosselin *et al.*, 2011). Identification of these factors is required to confirm these observations. It suggests that *Salmonella* express non-identified invasins able to mediate a Zipper process and factor(s) other than the T3SS-1 that induce Trigger-like invasion process(es). Type IV or type VI secretion systems are good candidates to induce Trigger-like cellular structures as they are able to translocate proteins directly into the host cell cytosol and as they are major virulence determinants involved in the pathogenesis of diverse Gram-negative bacteria (Oliveira *et al.*, 2006; Filloux *et al.*, 2008; Blondel *et al.*, 2010).

These observations thus open new avenues for identification of new invasion factors.

4. Conclusions and perspectives

Until recently, it was accepted that *Salmonella* enter cells only via its T3SS-1, which mediates a Trigger entry process. The T3SS-1-dependent invasion system has been widely described in the literature both at the bacterial and the cellular molecular levels. Moreover, the requirement of the T3SS-1 during intestinal and systemic infections has been demonstrated in some animals (Wallis & Galyov, 2000). However, an increasing number of reports describe that different serotypes of *Salmonella* can induce host infection without a functional T3SS-1 (Penheiter *et al.*, 1997; Jones *et al.*, 2001; Karasova *et al.*, 2010). This has been demonstrated with T3SS-1 mutants but also with clinical *Salmonella* strains (Hu *et al.*, 2008). However, the majority of *Salmonella* invasion system studies have focused on the T3SS-1 and we have little information concerning the T3SS-1-independent entry processes. Several invasins including PagN, Rck and HlyE have been recently identified in *Salmonella* and different investigations have provided evidences for other non-identified invasion factors (Aiastui *et al.*, 2010; Radtke *et al.*, 2010; Rosselin *et al.*, 2011; van Sorge *et al.*, 2011). In addition, the vast majority of this information has been obtained in a mouse model and with *S.* Typhimurium and much less data are available for other serotypes especially those adapted to pigs, cattle or poultry which represent major reservoirs of *Salmonella*. *Salmonella enterica* contain, over 2,500 diverse serotypes that have different host ranges, and cause diseases with severity ranging from subclinical colonization to serious systemic disease. Because the essential feature of the pathogenicity of *Salmonella* is its interaction with host cells, the identification of new entry routes could, in part, explain their different host ranges and disease symptoms.

The finding that *Salmonella* serotypes use different cell receptors and cell routes for host infection shows that the contribution of *Salmonella* genes to pathogenesis may be more complex than previously thought. These findings are changing our classical view of *Salmonella* pathogenicity. This new paradigm will modify the understanding of the mechanisms that lead to the different *Salmonella*-induced diseases and could allow us to revisit the host specificity bases.

5. References

Ahmer, B. M., van Reeuwijk, J., Timmers, C. D., Valentine, P. J. & Heffron, F. (1998). *Salmonella* typhimurium encodes an SdiA homolog, a putative quorum sensor of the LuxR family, that regulates genes on the virulence plasmid. *J Bacteriol*. Vol. 180, No. 5, pp. 1185-93.

Aiastui, A., Pucciarelli, M. G. & Garcia-del Portillo, F. (2010). *Salmonella* enterica serovar typhimurium invades fibroblasts by multiple routes differing from the entry into epithelial cells. *Infect Immun*. Vol. 78, No. 6, pp. 2700-13.

Akbar, S., Schechter, L. M., Lostroh, C. P. & Lee, C. A. (2003). AraC/XylS family members, HilD and HilC, directly activate virulence gene expression independently of HilA in *Salmonella* typhimurium. *Mol Microbiol*. Vol. 47, No. 3, pp. 715-28.

Akeda, Y. & Galan, J. E. (2004). Genetic analysis of the *Salmonella* enterica type III secretion-associated ATPase InvC defines discrete functional domains. *J Bacteriol*. Vol. 186, No. 8, pp. 2402-12.

Bakowski, M. A., Cirulis, J. T., Brown, N. F., Finlay, B. B. & Brumell, J. H. (2007). SopD acts cooperatively with SopB during *Salmonella* enterica serovar Typhimurium invasion. *Cell Microbiol*. Vol. 9, No. 12, pp. 2839-55.

Baumler, A. J., Tsolis, R. M., Bowe, F. A., Kusters, J. G., Hoffmann, S. & Heffron, F. (1996). The pef fimbrial operon of *Salmonella* typhimurium mediates adhesion to murine small intestine and is necessary for fluid accumulation in the infant mouse. *Infect Immun*. Vol. 64, No. 1, pp. 61-8.

Baxter, M. A., Fahlen, T. F., Wilson, R. L. & Jones, B. D. (2003). HilE interacts with HilD and negatively regulates hilA transcription and expression of the *Salmonella* enterica serovar Typhimurium invasive phenotype. *Infect Immun*. Vol. 71, No. 3, pp. 1295-305.

Baxter, M. A. & Jones, B. D. (2005). The fimYZ genes regulate *Salmonella* enterica Serovar Typhimurium invasion in addition to type 1 fimbrial expression and bacterial motility. *Infect Immun*. Vol. 73, No. 3, pp. 1377-85.

Belden, W. J. & Miller, S. I. (1994). Further characterization of the PhoP regulon: identification of new PhoP-activated virulence loci. *Infect Immun*. Vol. 62, No. 11, pp. 5095-101.

Blondel, C. J., Yang, H. J., Castro, B., Chiang, S., Toro, C. S., Zaldivar, M., Contreras, I., Andrews-Polymenis, H. L. & Santiviago, C. A. (2010). Contribution of the type VI secretion system encoded in SPI-19 to chicken colonization by *Salmonella* enterica serotypes Gallinarum and Enteritidis. *PLoS One*. Vol. 5, No. 7, pp. e11724.

Boonyom, R., Karavolos, M.H., Bulmer, D.M. & Khan, C.M. (2010). /Salmonella/ pathogenicity island 1 (SPI-1) type III secretion of SopD involves N- and C-terminal signals and direct binding to the InvC ATPase. Microbiology. vol.156, No.6, pp 1805-14.

Bouwman, C. W., Kohli, M., Killoran, A., Touchie, G. A., Kadner, R. J. & Martin, N. L. (2003). Characterization of SrgA, a *Salmonella* enterica serovar Typhimurium virulence plasmid-encoded paralogue of the disulfide oxidoreductase DsbA, essential for biogenesis of plasmid-encoded fimbriae. *J Bacteriol*. Vol. 185, No. 3, pp. 991-1000.

Bretscher, M. S. (1991). Lipid flow in locomoting cells. *Science*. Vol. 251, No. 4991, pp. 317-8.

Brown, M. D., Bry, L., Li, Z. & Sacks, D. B. (2007). IQGAP1 regulates *Salmonella* invasion through interactions with actin, Rac1, and Cdc42. *J Biol Chem*. Vol. 282, No. 41, pp. 30265-72.

Buchwald, G., Friebel, A., Galan, J. E., Hardt, W. D., Wittinghofer, A. & Scheffzek, K. (2002). Structural basis for the reversible activation of a Rho protein by the bacterial toxin SopE. *EMBO J.* Vol. 21, No. 13, pp. 3286-95.

Chang, J., Chen, J. & Zhou, D. (2005). Delineation and characterization of the actin nucleation and effector translocation activities of *Salmonella* SipC. *Mol Microbiol.* Vol. 55, No. 5, pp. 1379-89.

Chang, J., Myeni, S. K., Lin, T. L., Wu, C. C., Staiger, C. J. & Zhou, D. (2007). SipC multimerization promotes actin nucleation and contributes to *Salmonella*-induced inflammation. *Mol Microbiol.* Vol. 66, No. 6, pp. 1548-56.

Cirillo, D. M., Heffernan, E. J., Wu, L., Harwood, J., Fierer, J. & Guiney, D. G. (1996). Identification of a domain in Rck, a product of the *Salmonella* typhimurium virulence plasmid, required for both serum resistance and cell invasion. *Infect Immun.* Vol. 64, No. 6, pp. 2019-23.

Conner, C. P., Heithoff, D. M. & Mahan, M. J. (1998). In vivo gene expression: contributions to infection, virulence, and pathogenesis. *Curr Top Microbiol Immunol.* Vol. 225, No., pp. 1-12.

Coombes, B. K., Wickham, M. E., Lowden, M. J., Brown, N. F. & Finlay, B. B. (2005). Negative regulation of *Salmonella* pathogenicity island 2 is required for contextual control of virulence during typhoid. *Proc Natl Acad Sci U S A.* Vol. 102, No. 48, pp. 17460-5.

Cossart, P. & Sansonetti, P. J. (2004). Bacterial invasion: the paradigms of enteroinvasive pathogens. *Science.* Vol. 304, No. 5668, pp. 242-8.

Criss, A. K., Ahlgren, D. M., Jou, T. S., McCormick, B. A. & Casanova, J. E. (2001). The GTPase Rac1 selectively regulates *Salmonella* invasion at the apical plasma membrane of polarized epithelial cells. *J Cell Sci.* Vol. 114, No. Pt 7, pp. 1331-41.

Dai, S., Zhang, Y., Weimbs, T., Yaffe, M. B. & Zhou, D. (2007). Bacteria-generated PtdIns(3)P recruits VAMP8 to facilitate phagocytosis. *Traffic.* Vol. 8, No. 10, pp. 1365-74.

Darwin, K. H. & Miller, V. L. (1999). InvF is required for expression of genes encoding proteins secreted by the SPI1 type III secretion apparatus in *Salmonella* typhimurium. *J Bacteriol.* Vol. 181, No. 16, pp. 4949-54.

Dyszel, J. L., Smith, J. N., Lucas, D. E., Soares, J. A., Swearingen, M. C., Vross, M. A., Young, G. M. & Ahmer, B. M. (2010). *Salmonella* enterica serovar Typhimurium can detect acyl homoserine lactone production by Yersinia enterocolitica in mice. *J Bacteriol.* Vol. 192, No. 1, pp. 29-37.

Ellermeier, C. D. & Slauch, J. M. (2004). RtsA coordinately regulates DsbA and the *Salmonella* pathogenicity island 1 type III secretion system. *J Bacteriol.* Vol. 186, No. 1, pp. 68-79.

Ellermeier, J. R. & Slauch, J. M. (2007). Adaptation to the host environment: regulation of the SPI1 type III secretion system in *Salmonella* enterica serovar Typhimurium. *Curr Opin Microbiol.* Vol. 10, No. 1, pp. 24-9.

Eriksson, S., Lucchini, S., Thompson, A., Rhen, M. & Hinton, J. C. (2003). Unravelling the biology of macrophage infection by gene expression profiling of intracellular *Salmonella* enterica. *Mol Microbiol.* Vol. 47, No. 1, pp. 103-18.

Fagan, R. P., Lambert, M. A. & Smith, S. G. (2008). The hek outer membrane protein of Escherichia coli strain RS218 binds to proteoglycan and utilizes a single extracellular loop for adherence, invasion, and autoaggregation. *Infect Immun.* Vol. 76, No. 3, pp. 1135-42.

Fahlen, T. F., Mathur, N. & Jones, B. D. (2000). Identification and characterization of mutants with increased expression of hilA, the invasion gene transcriptional activator of *Salmonella* typhimurium. *FEMS Immunol Med Microbiol.* Vol. 28, No. 1, pp. 25-35.

Filloux, A., Hachani, A. & Bleves, S. (2008). The bacterial type VI secretion machine: yet another player for protein transport across membranes. *Microbiology*. Vol. 154, No. Pt 6, pp. 1570-83.

Folkesson, A., Advani, A., Sukupolvi, S., Pfeifer, J. D., Normark, S. & Lofdahl, S. (1999). Multiple insertions of fimbrial operons correlate with the evolution of *Salmonella* serovars responsible for human disease. *Mol Microbiol*. Vol. 33, No. 3, pp. 612-22.

Friebel, A., Ilchmann, H., Aepfelbacher, M., Ehrbar, K., Machleidt, W. & Hardt, W. D. (2001). SopE and SopE2 from *Salmonella* typhimurium activate different sets of RhoGTPases of the host cell. *J Biol Chem*. Vol. 276, No. 36, pp. 34035-40.

Fu, Y. & Galan, J. E. (1998). The *Salmonella* typhimurium tyrosine phosphatase SptP is translocated into host cells and disrupts the actin cytoskeleton. *Mol Microbiol*. Vol. 27, No. 2, pp. 359-68.

Fu, Y. & Galan, J. E. (1999). A *Salmonella* protein antagonizes Rac-1 and Cdc42 to mediate host-cell recovery after bacterial invasion. *Nature*. Vol. 401, No. 6750, pp. 293-7.

Fuentes, J. A., Villagra, N., Castillo-Ruiz, M. & Mora, G. C. (2008). The *Salmonella* Typhi hlyE gene plays a role in invasion of cultured epithelial cells and its functional transfer to S. Typhimurium promotes deep organ infection in mice. *Res Microbiol*. Vol. 159, No. 4, pp. 279-87.

Galan, J. E. & Wolf-Watz, H. (2006). Protein delivery into eukaryotic cells by type III secretion machines. *Nature*. Vol. 444, No. 7119, pp. 567-73.

Galkin, V. E., Orlova, A., VanLoock, M. S., Zhou, D., Galan, J. E. & Egelman, E. H. (2002). The bacterial protein SipA polymerizes G-actin and mimics muscle nebulin. *Nat Struct Biol*. Vol. 9, No. 7, pp. 518-21.

Garner, M. J., Hayward, R. D. & Koronakis, V. (2002). The *Salmonella* pathogenicity island 1 secretion system directs cellular cholesterol redistribution during mammalian cell entry and intracellular trafficking. *Cell Microbiol*. Vol. 4, No. 3, pp. 153-65.

Groisman, E. A. & Ochman, H. (1993). Cognate gene clusters govern invasion of host epithelial cells by *Salmonella* typhimurium and Shigella flexneri. *EMBO J*. Vol. 12, No. 10, pp. 3779-87.

Gunn, J. S., Alpuche-Aranda, C. M., Loomis, W. P., Belden, W. J. & Miller, S. I. (1995). Characterization of the *Salmonella* typhimurium pagC/pagD chromosomal region. *J Bacteriol*. Vol. 177, No. 17, pp. 5040-7.

Hanisch, J., Kolm, R., Wozniczka, M., Bumann, D., Rottner, K. & Stradal, T. E. (2011). Activation of a RhoA/myosin II-dependent but Arp2/3 complex-independent pathway facilitates *Salmonella* invasion. *Cell Host Microbe*. Vol. 9, No. 4, pp. 273-85.

Hartwig, J. H. & Kwiatkowski, D. J. (1991). Actin-binding proteins. *Curr Opin Cell Biol*. Vol. 3, No. 1, pp. 87-97.

Hayward, R. D., Cain, R. J., McGhie, E. J., Phillips, N., Garner, M. J. & Koronakis, V. (2005). Cholesterol binding by the bacterial type III translocon is essential for virulence effector delivery into mammalian cells. *Mol Microbiol*. Vol. 56, No. 3, pp. 590-603.

Hayward, R. D., McGhie, E. J. & Koronakis, V. (2000). Membrane fusion activity of purified SipB, a *Salmonella* surface protein essential for mammalian cell invasion. *Mol Microbiol*. Vol. 37, No. 4, pp. 727-39.

Heffernan, E. J., Harwood, J., Fierer, J. & Guiney, D. (1992). The *Salmonella* typhimurium virulence plasmid complement resistance gene rck is homologous to a family of virulence-related outer membrane protein genes, including pagC and ail. *J Bacteriol*. Vol. 174, No. 1, pp. 84-91.

Heithoff, D. M., Conner, C. P., Hentschel, U., Govantes, F., Hanna, P. C. & Mahan, M. J. (1999). Coordinate intracellular expression of *Salmonella* genes induced during infection. *J Bacteriol.* Vol. 181, No. 3, pp. 799-807.

Higashide, W., Dai, S., Hombs, V. P. & Zhou, D. (2002). Involvement of SipA in modulating actin dynamics during *Salmonella* invasion into cultured epithelial cells. *Cell Microbiol.* Vol. 4, No. 6, pp. 357-65.

Honda, A., Nogami, M., Yokozeki, T., Yamazaki, M., Nakamura, H., Watanabe, H., Kawamoto, K., Nakayama, K., Morris, A. J., Frohman, M. A. & Kanaho, Y. (1999). Phosphatidylinositol 4-phosphate 5-kinase alpha is a downstream effector of the small G protein ARF6 in membrane ruffle formation. *Cell.* Vol. 99, No. 5, pp. 521-32.

Hu, Q., Coburn, B., Deng, W., Li, Y., Shi, X., Lan, Q., Wang, B., Coombes, B. K. & Finlay, B. B. (2008). *Salmonella* enterica serovar Senftenberg human clinical isolates lacking SPI-1. *J Clin Microbiol.* Vol. 46, No. 4, pp. 1330-6.

Hueck, C. J. (1998). Type III protein secretion systems in bacterial pathogens of animals and plants. *Microbiol Mol Biol Rev.* Vol. 62, No. 2, pp. 379-433.

Hueck, C. J., Hantman, M. J., Bajaj, V., Johnston, C., Lee, C. A. & Miller, S. I. (1995). *Salmonella* typhimurium secreted invasion determinants are homologous to Shigella Ipa proteins. *Mol Microbiol.* Vol. 18, No. 3, pp. 479-90.

Huett, A., Leong, J. M., Podolsky, D. K. & Xavier, R. J. (2009). The cytoskeletal scaffold Shank3 is recruited to pathogen-induced actin rearrangements. *Exp Cell Res.* Vol. 315, No. 12, pp. 2001-11.

Ibarra, J. A. & Steele-Mortimer, O. (2009). *Salmonella*--the ultimate insider. *Salmonella* virulence factors that modulate intracellular survival. *Cell Microbiol.* Vol. 11, No. 11, pp. 1579-86.

Jones, M. A., Hulme, S. D., Barrow, P. A. & Wigley, P. (2007). The *Salmonella* pathogenicity island 1 and *Salmonella* pathogenicity island 2 type III secretion systems play a major role in pathogenesis of systemic disease and gastrointestinal tract colonization of *Salmonella* enterica serovar Typhimurium in the chicken. *Avian Pathol.* Vol. 36, No. 3, pp. 199-203.

Jones, M. A., Wigley, P., Page, K. L., Hulme, S. D. & Barrow, P. A. (2001). *Salmonella* enterica serovar Gallinarum requires the *Salmonella* pathogenicity island 2 type III secretion system but not the *Salmonella* pathogenicity island 1 type III secretion system for virulence in chickens. *Infect Immun.* Vol. 69, No. 9, pp. 5471-6.

Kaniga, K., Trollinger, D. & Galan, J. E. (1995). Identification of two targets of the type III protein secretion system encoded by the inv and spa loci of *Salmonella* typhimurium that have homology to the Shigella IpaD and IpaA proteins. *J Bacteriol.* Vol. 177, No. 24, pp. 7078-85.

Karasova, D., Sebkova, A., Havlickova, H., Sisak, F., Volf, J., Faldyna, M., Ondrackova, P., Kummer, V. & Rychlik, I. (2010). Influence of 5 major *Salmonella* pathogenicity islands on NK cell depletion in mice infected with *Salmonella* enterica serovar Enteritidis. *BMC Microbiol.* Vol. 10, No., pp. 75.

Kimbrough, T. G. & Miller, S. I. (2000). Contribution of *Salmonella* typhimurium type III secretion components to needle complex formation. *Proc Natl Acad Sci U S A.* Vol. 97, No. 20, pp. 11008-13.

Kimbrough, T. G. & Miller, S. I. (2002). Assembly of the type III secretion needle complex of *Salmonella* typhimurium. *Microbes Infect.* Vol. 4, No. 1, pp. 75-82.

Kubori, T. & Galan, J. E. (2003). Temporal regulation of *Salmonella* virulence effector function by proteasome-dependent protein degradation. *Cell.* Vol. 115, No. 3, pp. 333-42.

Lambert, M. A. & Smith, S. G. (2008). The PagN protein of *Salmonella* enterica serovar Typhimurium is an adhesin and invasin. *BMC Microbiol.* Vol. 8, No., pp. 142.

Lambert, M. A. & Smith, S. G. (2009). The PagN protein mediates invasion via interaction with proteoglycan. *FEMS Microbiol Lett.* Vol. 297, No. 2, pp. 209-16.

Lara-Tejero, M., Kato, J., Wagner, S., Liu, X. & Galan, J. E. (2011). A sorting platform determines the order of protein secretion in bacterial type III systems. *Science.* Vol. 331, No. 6021, pp. 1188-91.

Ledeboer, N. A., Frye, J. G., McClelland, M. & Jones, B. D. (2006). *Salmonella* enterica serovar Typhimurium requires the Lpf, Pef, and Tafi fimbriae for biofilm formation on HEp-2 tissue culture cells and chicken intestinal epithelium. *Infect Immun.* Vol. 74, No. 6, pp. 3156-69.

Ly, K. T. & Casanova, J. E. (2009). Abelson tyrosine kinase facilitates *Salmonella* enterica serovar Typhimurium entry into epithelial cells. *Infect Immun.* Vol. 77, No. 1, pp. 60-9.

Mammarappallil, J. G. & Elsinghorst, E. A. (2000). Epithelial cell adherence mediated by the enterotoxigenic Escherichia coli tia protein. *Infect Immun.* Vol. 68, No. 12, pp. 6595-601.

Mattei, P. J., Faudry, E., Job, V., Izore, T., Attree, I. & Dessen, A. (2011). Membrane targeting and pore formation by the type III secretion system translocon. *FEBS J.* Vol. 278, No. 3, pp. 414-26.

McGhie, E. J., Hayward, R. D. & Koronakis, V. (2001). Cooperation between actin-binding proteins of invasive *Salmonella*: SipA potentiates SipC nucleation and bundling of actin. *EMBO J.* Vol. 20, No. 9, pp. 2131-9.

McGhie, E. J., Hayward, R. D. & Koronakis, V. (2004). Control of actin turnover by a *Salmonella* invasion protein. *Mol Cell.* Vol. 13, No. 4, pp. 497-510.

Mecsas, J., Welch, R., Erickson, J. W. & Gross, C. A. (1995). Identification and characterization of an outer membrane protein, OmpX, in Escherichia coli that is homologous to a family of outer membrane proteins including Ail of Yersinia enterocolitica. *J Bacteriol.* Vol. 177, No. 3, pp. 799-804.

Michael, B., Smith, J. N., Swift, S., Heffron, F. & Ahmer, B. M. (2001). SdiA of *Salmonella* enterica is a LuxR homolog that detects mixed microbial communities. *J Bacteriol.* Vol. 183, No. 19, pp. 5733-42.

Mijouin, L., Rosselin, M., Bottreau, E., Pizarro-Cerda, J., Cossart, P., Velge, P. & Wiedemann A (2012). /*Salmonella*/ Enteritidis Rck-mediated invasion requires activation of Rac1 dependent of Class I PI 3-kinases-Akt signalling pathway. /FASEB J./ vol.26, No.4, pp 1569-81.

Miki, T., Okada, N., Shimada, Y. & Danbara, H. (2004). Characterization of *Salmonella* pathogenicity island 1 type III secretion-dependent hemolytic activity in *Salmonella* enterica serovar Typhimurium. *Microb Pathog.* Vol. 37, No. 2, pp. 65-72.

Miller, S. I., Kukral, A. M. & Mekalanos, J. J. (1989). A two-component regulatory system (phoP phoQ) controls *Salmonella* typhimurium virulence. *Proc Natl Acad Sci U S A.* Vol. 86, No. 13, pp. 5054-8.

Miller, V. L., Beer, K. B., Heusipp, G., Young, B. M. & Wachtel, M. R. (2001). Identification of regions of Ail required for the invasion and serum resistance phenotypes. *Mol Microbiol.* Vol. 41, No. 5, pp. 1053-62.

Misselwitz, B., Dilling, S., Vonaesch, P., Sacher, R., Snijder, B., Schlumberger, M., Rout, S., Stark, M., von Mering, C., Pelkmans, L. & Hardt, W. D. (2011). RNAi screen of *Salmonella* invasion shows role of COPI in membrane targeting of cholesterol and Cdc42. *Mol Syst Biol.* Vol. 7, No., pp. 474.

Mullins, R. D., Heuser, J. A. & Pollard, T. D. (1998). The interaction of Arp2/3 complex with actin: nucleation, high affinity pointed end capping, and formation of branching networks of filaments. *Proc Natl Acad Sci U S A*. Vol. 95, No. 11, pp. 6181-6.

Myeni, S. K. & Zhou, D. (2010). The C terminus of SipC binds and bundles F-actin to promote *Salmonella* invasion. *J Biol Chem*. Vol. 285, No. 18, pp. 13357-63.

Nichols, C. D. & Casanova, J. E. (2010). *Salmonella*-directed recruitment of new membrane to invasion foci via the host exocyst complex. *Curr Biol*. Vol. 20, No. 14, pp. 1316-20.

Norris, F. A., Wilson, M. P., Wallis, T. S., Galyov, E. E. & Majerus, P. W. (1998). SopB, a protein required for virulence of *Salmonella* dublin, is an inositol phosphate phosphatase. *Proc Natl Acad Sci U S A*. Vol. 95, No. 24, pp. 14057-9.

Oliveira, M. J., Costa, A. C., Costa, A. M., Henriques, L., Suriano, G., Atherton, J. C., Machado, J. C., Carneiro, F., Seruca, R., Mareel, M., Leroy, A. & Figueiredo, C. (2006). Helicobacter pylori induces gastric epithelial cell invasion in a c-Met and type IV secretion system-dependent manner. *J Biol Chem*. Vol. 281, No. 46, pp. 34888-96.

Patel, J. C. & Galan, J. E. (2005). Manipulation of the host actin cytoskeleton by *Salmonella*-- all in the name of entry. *Curr Opin Microbiol*. Vol. 8, No. 1, pp. 10-5.

Patel, J. C. & Galan, J. E. (2006). Differential activation and function of Rho GTPases during *Salmonella*-host cell interactions. *J Cell Biol*. Vol. 175, No. 3, pp. 453-63.

Penheiter, K. L., Mathur, N., Giles, D., Fahlen, T. & Jones, B. D. (1997). Non-invasive *Salmonella* typhimurium mutants are avirulent because of an inability to enter and destroy M cells of ileal Peyer's patches. *Mol Microbiol*. Vol. 24, No. 4, pp. 697-709.

Perrett, C. A. & Jepson, M. A. (2009). Regulation of *Salmonella*-induced membrane ruffling by SipA differs in strains lacking other effectors. *Cell Microbiol*. Vol. 11, No. 3, pp. 475-87.

Pollard, T. D., Blanchoin, L. & Mullins, R. D. (2000). Molecular mechanisms controlling actin filament dynamics in nonmuscle cells. *Annu Rev Biophys Biomol Struct*. Vol. 29, No., pp. 545-76.

Queiroz, M. H., Madrid, C., Paytubi, S., Balsalobre, C. & Juarez, A. (2011). Integration Host Factor alleviates H-NS silencing of the *Salmonella* enterica serovar Typhimurium master regulator of SPI1, hilA. *Microbiology*. Vol. No.

Radtke, A. L., Wilson, J. W., Sarker, S. & Nickerson, C. A. (2010). Analysis of interactions of *Salmonella* type three secretion mutants with 3-D intestinal epithelial cells. *PLoS One*. Vol. 5, No. 12, pp. e15750.

Rathinavelan, T., Tang, C. & De Guzman, R. N. (2011). Characterization of the interaction between the *Salmonella* type III secretion system tip protein SipD and the needle protein PrgI by paramagnetic relaxation enhancement. *J Biol Chem*. Vol. 286, No. 6, pp. 4922-30.

Reeve, J. N. & Shaw, J. E. (1979). Lambda encodes an outer membrane protein: the lom gene. *Mol Gen Genet*. Vol. 172, No. 3, pp. 243-8.

Rittmeyer, E. N., Daniel, S., Hsu, S. C. & Osman, M. A. (2008). A dual role for IQGAP1 in regulating exocytosis. *J Cell Sci*. Vol. 121, No. Pt 3, pp. 391-403.

Rosselin, M., Abed, N., Virlogeux-Payant, I., Bottreau, E., Sizaret, P. Y., Velge, P. & Wiedemann, A. (2011). Heterogeneity of type III secretion system (T3SS)-1-independent entry mechanisms used by *Salmonella* Enteritidis to invade different cell types. *Microbiology*. Vol. 157, No. Pt 3, pp. 839-47.

Rosselin, M., Virlogeux-Payant, I., Roy, C., Bottreau, E., Sizaret, P. Y., Mijouin, L., Germon, P., Caron, E., Velge, P. & Wiedemann, A. (2010). Rck of *Salmonella* enterica, subspecies enterica serovar enteritidis, mediates zipper-like internalization. *Cell Res*. Vol. 20, No. 6, pp. 647-64.

Rotger, R. & Casadesus, J. (1999). The virulence plasmids of *Salmonella*. *Int Microbiol*. Vol. 2, No. 3, pp. 177-84.

Samudrala, R., Heffron, F. & McDermott, J. E. (2009). Accurate prediction of secreted substrates and identification of a conserved putative secretion signal for type III secretion systems. *PLoS Pathog*. Vol. 5, No. 4, pp. e1000375.

Scherer, C. A., Cooper, E. & Miller, S. I. (2000). The *Salmonella* type III secretion translocon protein SspC is inserted into the epithelial cell plasma membrane upon infection. *Mol Microbiol*. Vol. 37, No. 5, pp. 1133-45.

Schlumberger, M. C., Friebel, A., Buchwald, G., Scheffzek, K., Wittinghofer, A. & Hardt, W. D. (2003). Amino acids of the bacterial toxin SopE involved in G nucleotide exchange on Cdc42. *J Biol Chem*. Vol. 278, No. 29, pp. 27149-59.

Schraidt, O. & Marlovits, T. C. (2011). Three-dimensional model of *Salmonella*'s needle complex at subnanometer resolution. *Science*. Vol. 331, No. 6021, pp. 1192-5.

Schroeder, G. N. & Hilbi, H. (2008). Molecular pathogenesis of Shigella spp.: controlling host cell signaling, invasion, and death by type III secretion. *Clin Microbiol Rev*. Vol. 21, No. 1, pp. 134-56.

Sechi, A. S. & Wehland, J. (2000). The actin cytoskeleton and plasma membrane connection: PtdIns(4,5)P(2) influences cytoskeletal protein activity at the plasma membrane. *J Cell Sci*. Vol. 113 Pt 21, No., pp. 3685-95.

Shi, J. & Casanova, J. E. (2006). Invasion of host cells by *Salmonella* typhimurium requires focal adhesion kinase and p130Cas. *Mol Biol Cell*. Vol. 17, No. 11, pp. 4698-708.

Smith, J. N. & Ahmer, B. M. (2003). Detection of other microbial species by *Salmonella*: expression of the SdiA regulon. *J Bacteriol*. Vol. 185, No. 4, pp. 1357-66.

Smith, J. N., Dyszel, J. L., Soares, J. A., Ellermeier, C. D., Altier, C., Lawhon, S. D., Adams, L. G., Konjufca, V., Curtiss, R., 3rd, Slauch, J. M. & Ahmer, B. M. (2008). SdiA, an N-acylhomoserine lactone receptor, becomes active during the transit of *Salmonella* enterica through the gastrointestinal tract of turtles. *PLoS One*. Vol. 3, No. 7, pp. e2826.

Stebbins, C. E. & Galan, J. E. (2000). Modulation of host signaling by a bacterial mimic: structure of the *Salmonella* effector SptP bound to Rac1. *Mol Cell*. Vol. 6, No. 6, pp. 1449-60.

Takeuchi, A. (1967). Electron microscope studies of experimental *Salmonella* infection. I. Penetration into the intestinal epithelium by *Salmonella* typhimurium. *Am J Pathol*. Vol. 50, No. 1, pp. 109-36.

Tampakaki, A. P., Fadouloglou, V. E., Gazi, A. D., Panopoulos, N. J. & Kokkinidis, M. (2004). Conserved features of type III secretion. *Cell Microbiol*. Vol. 6, No. 9, pp. 805-16.

Terebiznik, M. R., Vieira, O. V., Marcus, S. L., Slade, A., Yip, C. M., Trimble, W. S., Meyer, T., Finlay, B. B. & Grinstein, S. (2002). Elimination of host cell PtdIns(4,5)P(2) by bacterial SigD promotes membrane fission during invasion by *Salmonella*. *Nat Cell Biol*. Vol. 4, No. 10, pp. 766-73.

van Sorge, A. J., Termote, J. U., de Vries, M. J., Boonstra, F. N., Stellingwerf, C. & Schalij-Delfos, N. E. (2011). The incidence of visual impairment due to retinopathy of prematurity (ROP) and concomitant disabilities in the Netherlands: a 30 year overview. *Br J Ophthalmol*. Vol. 95, No. 7, pp. 937-41.

Vazquez-Torres, A., Jones-Carson, J., Baumler, A. J., Falkow, S., Valdivia, R., Brown, W., Le, M., Berggren, R., Parks, W. T. & Fang, F. C. (1999). Extraintestinal dissemination of *Salmonella* by CD18-expressing phagocytes. *Nature*. Vol. 401, No. 6755, pp. 804-8.

Velge, P., Cloeckaert, A. & Barrow, P. (2005). Emergence of *Salmonella* epidemics: the problems related to *Salmonella* enterica serotype Enteritidis and multiple antibiotic resistance in other major serotypes. *Vet Res.* Vol. 36, No. 3, pp. 267-88.

Vivero, A., Banos, R. C., Mariscotti, J. F., Oliveros, J. C., Garcia-del Portillo, F., Juarez, A. & Madrid, C. (2008). Modulation of horizontally acquired genes by the Hha-YdgT proteins in *Salmonella* enterica serovar Typhimurium. *J Bacteriol.* Vol. 190, No. 3, pp. 1152-6.

Wai, S. N., Lindmark, B., Soderblom, T., Takade, A., Westermark, M., Oscarsson, J., Jass, J., Richter-Dahlfors, A., Mizunoe, Y. & Uhlin, B. E. (2003). Vesicle-mediated export and assembly of pore-forming oligomers of the enterobacterial ClyA cytotoxin. *Cell.* Vol. 115, No. 1, pp. 25-35.

Wallar, L. E., Bysice, A. M. & Coombes, B. K. (2011). The non-motile phenotype of *Salmonella* hha ydgT mutants is mediated through PefI-SrgD. *BMC Microbiol.* Vol. 11, No., pp. 141.

Wallis, T. S. & Galyov, E. E. (2000). Molecular basis of *Salmonella*-induced enteritis. *Mol Microbiol.* Vol. 36, No. 5, pp. 997-1005.

Watson, K. G. & Holden, D. W. (2010). Dynamics of growth and dissemination of *Salmonella* in vivo. *Cell Microbiol.* Vol. 12, No. 10, pp. 1389-97.

Winnen, B., Schlumberger, M. C., Sturm, A., Schupbach, K., Siebenmann, S., Jenny, P. & Hardt, W. D. (2008). Hierarchical effector protein transport by the *Salmonella* Typhimurium SPI-1 type III secretion system. *PLoS One.* Vol. 3, No. 5, pp. e2178.

Wozniak, C. E., Lee, C. & Hughes, K. T. (2009). T-POP array identifies EcnR and PefI-SrgD as novel regulators of flagellar gene expression. *J Bacteriol.* Vol. 191, No. 5, pp. 1498-508.

Yao, Y., Dickerson, T. J., Hixon, M. S. & Dyson, H. J. (2007). NMR detection of adventitious xylose binding to the quorum-sensing protein SdiA of Escherichia coli. *Bioorg Med Chem Lett.* Vol. 17, No. 22, pp. 6202-5.

Zhou, D., Chen, L. M., Hernandez, L., Shears, S. B. & Galan, J. E. (2001). A *Salmonella* inositol polyphosphatase acts in conjunction with other bacterial effectors to promote host cell actin cytoskeleton rearrangements and bacterial internalization. *Mol Microbiol.* Vol. 39, No. 2, pp. 248-59.

Zhou, D., Mooseker, M. S. & Galan, J. E. (1999a). An invasion-associated *Salmonella* protein modulates the actin-bundling activity of plastin. *Proc Natl Acad Sci U S A.* Vol. 96, No. 18, pp. 10176-81.

Zhou, D., Mooseker, M. S. & Galan, J. E. (1999b). Role of the S. typhimurium actin-binding protein SipA in bacterial internalization. *Science.* Vol. 283, No. 5410, pp. 2092-5.

3

Molecular Diagnosis of Enteric Fever: Progress and Perspectives

Liqing Zhou, Thomas Darton, Claire Waddington and Andrew J. Pollard
University of Oxford
United Kingdom

1. Introduction

Enteric fever is a severe systemic Gram-negative bacterial infection caused by several serovars of *Salmonella enterica* subspecies *enterica*, including *S.* Typhi and *S.* Paratyphi serotypes A (most commonly), B and C. It is characterised by high fever and a myriad of other non-specific features, including abdominal pain and constipation, headache, myalgia and arthralgia, cough, lymphadenopathy and rash. *S.* Typhi, the human-specific causative agent of typhoid fever, is thought to account for an estimated 21 million new cases and 216,000 deaths every year (Crump et al., 2004). *S.* Typhi is generally transmitted in food and water contaminated with faeces from those excreting bacteria, either during the acute illness or during chronic asymptomatic carriage, although infection of health-care or laboratory workers through poor hygiene practices or accidental exposure is also described. Transmission in regions with adequate sanitation and sewage facilities is uncommon as, in general, a relatively high inoculum is required to survive the gastric acid environment and cause infection. Enteric fever is therefore most common in resource-poor settings where the provision of clean drinking water and sewage disposal facilities is inadequate. South and Central Asia, Africa and South and Central America are considered endemic for this disease and particularly high incidence rates are found in the Indian sub-continent and South-east Asia, with rates exceeding 100 per 100,000 population per year (Bhan et al., 2005). In other countries typhoid fever remains an important consideration for travellers both pre- and post-travel (Levine et al., 1982; Ackers et al., 2000; Bhan et al., 2005).

The accurate and rapid clinical diagnosis of enteric fever in these regions is obfuscated by the range of other common fever-causing infections including malaria, dengue fever, leptospirosis, melioidosis and the rickettsioses. Accurate diagnosis to differentiate typhoid fever from these conditions is often difficult, both in the clinic and in the laboratory, but is imperative for effective treatment selection. Even in highly-resourced western countries, physicians often start typhoid treatment empirically whilst awaiting confirmation of the diagnosis. Treatment decisions are further complicated by the increasing prevalence of antibiotic resistance amongst clinical isolates due to plasmid-mediated multidrug resistance (in particular the gyrA gene mutation, conferring variable fluoroquinolone resistance in both *S.* Typhi and *S.* Paratyphi A (Chau et al., 2007)) and the potential for extended-spectrum β-lactamase (ESBL) and carbapenemase-producing strains (Al Naiemi et al., 2008; Pokharel et al., 2006; Nordmann et al., 2008). Rates of illness caused by *S.* Parathyphi and non–typhoidal *Salmonella* are increasing in many endemic areas further complicating accurate laboratory testing (Ochiai et al., 2005; Palit et al., 2006).

It has long been accepted that vaccines represent the most cost effective approach to control typhoid infection, especially in the era of widespread and increasing antibiotic resistance (Parry et al., 2002; Whitaker et al., 2009). However, few countries have taken up routine typhoid immunization, partially due to uncertainty on disease burden and vaccine effectiveness. The development of cheap and reliable enteric fever diagnostics would play a key role in more accurately defining the scale of the problem and thus facilitating both long-term disease control and individual patient treatment (Baker et al., 2010). A combination of accurate diagnosis, effective vaccination and directed treatment could ultimately lead to the eradication of this human-restricted infection if appropriately implemented. Here we review the current means available for enteric fever diagnosis and the progress being made in improving molecular diagnostics in particular.

2. Clinical diagnosis of enteric fever

Enteric fever may affect individuals of any age; recently it has been shown to affect a much higher proportion of children aged less than 5 years than previously thought, causing a similar range of signs and symptoms to those seen in adults (Sinha et al., 1999). Immunosuppressed individuals, those with reduced gastric acid production, biliary and urinary tract abnormalities, haemoglobinpathies and other concomitant infectious diseases (including malaria and schistosomiasis) are at higher risk of acquiring infection and at risk of developing more severe or disseminated disease (Gotuzzo et al., 1991; Khosla et al., 1993; Mathai et al., 1995; Bhan et al., 2002; Crawford et al., 2010).

The clinical presentation of typhoid fever is notoriously variable, ranging from non-specific fever symptoms to fulminant Gram-negative sepsis with multisystem disease. The incubation period is classically 10 to 14 days although can range from 5 to 21 days. Early evidence suggested that as well as asymptomatic carriers, some individuals are capable of remaining asymptomatic and afebrile despite demonstrable bacteraemia (Snyder et al., 1963). The incubation period is likely to be directly proportional to the inoculum ingested and the cell-mediated immune response of the individual infected, although precise correlates of protection have yet to be determined (Sztein, 2007).

In the early days following infection, individuals may develop diarrhoea and abdominal discomfort. Diarrhoea is thought to be more common in certain geographic areas and in individuals with HIV/AIDS and in children less than 1 year of age (Butler et al., 1991). After a variable asymptomatic duration, individuals may develop constipation (10-38%), abdominal pain (30-40%), headache (often a dull frontal aching, 62%) and fever (Stuart & Pollen, 1946; Clark et al., 2010). Various studies have shown that fever is present in from 75 to 100% of microbiologically-confirmed cases on presentation (Stuart & Pullen, 1946; Butler et al., 1991; Clark et al., 2010); it classically starts low and increases in a saw-toothed pattern, often to between 39 and 40°C by the second week (see figure 1).

The spectrum of symptoms experienced is highly varied, and therefore the diagnosis may be missed particularly in areas where other febrile illnesses, such as malaria, tuberculosis or dengue, are common. Other presentations may include a more 'food poisoning' type illness with diarrhoea and vomiting or a predominantly respiratory presentation with symptoms including cough and audible crackles on chest auscultation. Other clinical findings of note include a relative bradycardia (Faget's sign, which occurs in less than 50% of patients), hepatosplenomegaly (20 to 50%) and Rose spots (up to 25%), which are classically described as salmon pink evanescent maculopapular spots seen towards the end of the first week of illness on the trunk, and from which *S.* Typhi may be cultured if biopsied (Parry et al., 2002; World Health Organization, 2003).

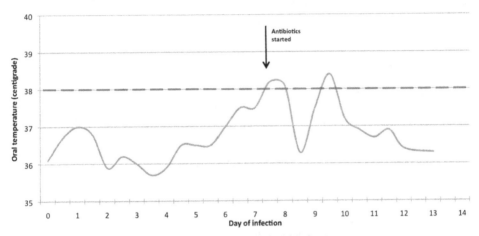

Fig. 1. The variation of oral temperature during typhoid infection

Presentation of neonatal typhoid fever resulting from vertical transmission during late pregnancy is usually within 3 days of delivery; signs including fever, vomiting, diarrhoea, and abdominal distension (Bhan et al., 2005). Significant hepatomegaly and jaundice and seizures can occur (Butler et al., 1991). Typhoid fever typically presents as a milder or atypical illness, often as a severe pneumonia, in children younger than 5 years (Mahle & Levine, 1993). The rate of severe complications is lower than in older age-groups (Mahle & Levine, 1993; Chiu & Lin, 1999; Sinha et al., 1999; Bhan et al., 2005).

Duration of illness before therapy, choice of antimicrobial therapy, strain virulence, inoculum size, previous exposure or vaccination, and other host factors such as HLA type, AIDS or other immune suppression, antacid consumption or concomittant *H. pylori* infection (Bhan et al., 2002) affect severity of the disease. Depending on the clinical resources available, approximately 10–15% of patients may develop more severe disease charaterised by the development of abdominal complications (Bhan et al., 2005). Gastrointestinal bleeding, intestinal perforation and typhoid encephalopathy are the commonest complications (Ali et al., 1997; Parry et al., 2002; World Health Organization, 2003; Bhan et al., 2005). The more details of clinical features of typhoid complications are described in the Seminar by Bhan et al. (Bhan et al., 2005).

Traditionally, the clinical features of paratyphoid fever were thought to be similar or milder than those of typhoid fever. With increasing incidence and more data now available, studies have started to demonstrate an equivalent or even increased rate of complications with paratyphoid infections (Ekdahl et al., 2005; Meltzer et al., 2005; Vollaard et al., 2005; Maskey et al., 2006; Woods et al., 2006). *S.* Paratyphi A, B or C may present with either systemic (Lee et al., 2000; Rajagopal et al., 2002; Mohanty et al., 2003) or localised infection (Fangtham et al., 2008). A relapse rate of 8% has been reported with S. Paratyphi A which is increasing in incidence throughout Southeast Asia (Ochiai et al., 2005; Woods et al., 2006; Fangtham et al., 2008) and may be associated with higher rates of complicated disease and outbreaks of infection (Khan et al., 2007; Pandit et al., 2008; Patel et al., 2010). S. Paratyphi A and B may present with a non-specific Salmonella gastroenteritis with diarrhoea being a predominant symptom (Thisyakorn et al., 1987; Yang et al., 2010). Gastrointestinal symptoms are usually not present with S. Paratyphi C infection but there have been cases with systemic complications such as septicaemia and arthritis (Lang et al., 1992).

3. Laboratory diagnosis of enteric fever

Current widely used methods for the diagnosis of individuals with enteric fever include bacterial culture, microscopy and serological assays, specifically the Widal test, which have been recently reviewed by Bhan et al. (2005), Bhutta (2006), Kundu et al. (2006), Wain & Hosoglu (2008) and Parry et al. (2011). Molecular diagnostics of enteric fever, in particular nucleic acid amplification by polymerase chain reaction (PCR), have been growing rapidly in last decade although they are confined within the research setting.

3.1 Bacterial culture

Accurate diagnosis of enteric fever requires isolation (or detection) of the causative organism, preferably from a sterile site (World Health Organization, 2003). Even though an array of specimens including whole blood, bone marrow, stool, duodenal fluid, urine and skin (Rose spots) (Gilman et al., 1975; Vallenas et al., 1985; Hoffman et al., 1986; Rubin et al., 1990) have historically been shown to harbor cultivable bacteria, blood is the most common specimen submitted for culture of S. Typhi (Parry et al., 2002; Wain and Hosoglu 2008). Between 45 and 70% of patients with typhoid fever may be diagnosed by blood culture (World Health Organization, 2003; Wain et al., 2001, 2008). The sensitivity of culture from blood is dependent on a variety of factors including the volume of blood taken (and its ratio to enrichment broth), pre-treatment with antibiotics and delay in transportation of the sample to the laboratory (Wain et al., 2008). As the number of circulating bacteria may be extremely low and predominantly intracellular (over 50% in one study (Wain et al., 2001)), any of these variables may significantly affect the growth and therefore the isolation rate. Use of selective media such as ox bile broth may increase this rate, as, while selective for bile resistant organisms, it inhibits some of the bactericidal activity of blood and is capable of releasing intracellular bacteria (Coleman & Buxton, 1907; Kaye et al., 1966; Wain et al., 2008). Research performed in our laboratory has also confirmed that bile (as ox bile soy tryptone broth) causes selective lysis of mammalian cells whilst leaving bacterial cells intact and capable of unhindered growth in culture (Zhou & Pollard, 2010). Whilst useful for research settings, selective culture of blood in bile-containing media outside of highly endemic regions is unhelpful in the general microbiology laboratory although alternative additives such as saponin have also been investigated (Murray et al., 1991; Wain & Hosoglu 2008; Wain et al., 2008).

Although it is thought that a significant inoculum is required to cause typhoid fever, in those with enhanced susceptibility, ingestion of even a small number of S. Typhi organisms may be sufficient to cause infection. Previous studies using a typhoid challenge model in healthy adult volunteers demonstrated that as few as 10^5 organisms were capable of causing disease following gastric acid suppression using milk (Glynn et al., 1995). In ongoing challenge studies, we have demonstrated that as few as 700 colony forming units (CFU) of non-attenuated live S. Typhi may cause clinical illness after gastric acid suppression using sodium bicarbonate. That very low numbers of S. Typhi are found circulating in the bloodstream at onset of symptoms in most typhoid cases is therefore not surprising; in 81 patients diagnosed with typhoid fever, a median level of 0.3 (IQR, 0.1-10) bacteria per millilitre of blood was found (Wain et al., 2001). Therefore, one of the key issues in typhoid diagnostics is how to detect the extremely low level bacteraemia present in a sick patient. Even using modern PCR and related diagnostics, current studies often still employ a pre-culture stage in order to try and maximise the organism detection rate (Nga et al., 2010; Zhou & Pollard, 2010).

Bone marrow harbors over 10 times as many organisms per unit volume than in the blood (Wain et al., 2008). Aside from the degree of patient discomfort involved, bone marrow aspiration and culture may therefore represent a useful addition to blood culture if appropriate facilities exist, particularly in patients who have been heavily antibiotic exposed (Wain et al., 2001) or who are being investigated for haematological conditions or pyrexia of unknown origin simultaneously (Volk et al., 1998).

Stool specimens are commonly collected during the diagnostic work-up of patients with typhoid infection, but there may be difficulty in obtaining specimens due to constipation when rectal swabs are a less good alternative. Stool should be cultured in selenite enrichment broth to maximise the culture yield (Moriñigo et al., 1993) for which standard selenite F medium appears at least as effective as selenite supplemented with mannitol (selenite M) (Wain et al., 2008). The results of a positive stool culture need to be interpreted in light of the clinical condition of the patient to exclude healthy carriers (such as `Typhoid Mary') (Soper, 1939). Stool cultures obtained from acutely ill patients may become positive before blood cultures, immediately preceeding either the primary or secondary bacteraemic phase, and their sensitivity increases with the quantity obtained (Personal observations; Wain et al., 2008). Stool cultures are therefore a useful aid to diagnosis and to guide public health prevention activities in certain settings.

Rose spot skin biopsies (Gilman et al., 1975; Wain et al., 1998) and urine samples may also be used for culturing S. Typhi, the latter being culture positive in approximately 7% of confirmed cases (Gilman et al., 1975). Duodenal contents obtained using a duodenal string test or aspiration may be more useful for culture identification of causative organisms, but the procedures required are often poorly tolerated, particularly by young children (Vallenas et al., 1985).

Most diagnoses of enteric fever are still made by blood culture followed by microbiological identification. However, blood culture, whilst considered "routine" in most resource-rich settings, is expensive, requiring specialist facilities and personnel, and time-consuming, taking at least 2 to 5 days for organism growth and positive identification.

3.2 Serological tests

Several serological tests have been developed in order to detect the presence of either S. Typhi antigens or the antibody response to it. The classic Widal test, a tube agglutination test developed by Widal F. in 1896 (Widal et al., 1896), detects the presence of agglutinating antibodies in the serum of infected/exposed patients against lipopolysaccharide (LPS; O) and flagella (H) antigens of *S.* Typhi (Olopoenia & King, 2000; World Health Organization, 2003). These antibodies present at 6 to 8 days and 10 to 12 days respectively, following infection; a 4-fold rise in either of these antibodies between acute and convalescent sera is diagnostic (World Health Organization, 2003). The test is only moderately specific for typhoid infection; however, studies from several areas, predominantly endemic for typhoid infection, demonstrate a significant variation in assay performance particularly when using a single Widal test result to make a typhoid fever diagnosis. Reasons for false-positive test results may include previous vaccination or exposure to natural infection, cross-reactivity with epitopes from other enterobacteriaceae or concomittant infections including malaria, typhus and other causes of bacteraemia (Reynolda et al., 1970; Levine et al., 1978; Olopoenia & King, 2000; House et al., 2001; World Health Organization, 2003; Omuse et al., 2010). Likewise, false-negative tests are also seen which may be due to previous antibiotic exposure or other medical conditions capable of reducing the antibody response generated. Widal tests are relatively inexpensive however, particularly in comparison to bacterial culture methods, and are therefore still widely used (Bakr et al., 2011) and are possibly of more benefit in non-endemic settings (Levine et al., 1978; Chew et al., 1992).

Much effort has been put into improving on the classic Widal test over the last twenty years specifically in order to improve the speed and reliability of serological testing (Bhutta & Mansurali, 1999; House et al., 2001; Gasem et al., 2002; Hatta et al., 2002; Jesudason et al., 2002; Olsen et al., 2004; Tam et al., 2008; Fadeel et al., 2011). Several of these assays have subsequently become commerically available; Typhidot® (Malaysian Biodiagnostic Research SDN BHD, Malaysia) and TUBEX assays (IDL Bideh, Solletuna, Sweden) are discussed in further detail below.

Typhidot® is a dot enzyme-linked immunosorbent assay capable of detecting both IgM and IgG antibodies against a *S.* Typhi-specific 50kDa outer membrane protein (OMP) (Ismail et al., 1991; Choo et al., 1994, 1999). OMP dotted onto a nitrocellulose strip is probed with test sera and developed using peroxidase-conjugated antihuman IgM/IgG antibodies and a substrate for colour development (Choo et al., 1994; Kawano et al., 2007).

TUBEX-TF® is an inhibition binding assay that detects the presence of the O9 component of *S.* Typhi LPS. Binding of *S.* Typhi LPS (O9) antibody-coated indicator to *S.* Typhi LPS (antigen)-coated magnetic particles is inhibited by patient sera containing anti-O9 antibodies, which results in a quantitative red-blue colour change (Lim et al., 1998; Oracz et al., 2003). Elevated levels of anti-O9 IgM antibodies together with typical clinical symptoms of typhoid fever probably indicates acute infection with *S.* Typhi (Tam & Lim, 2003; Feleszko et al., 2004; Tam et al., 2008). Subsequent modification of the antigens used has resulted in a similar test for paratyphoid fever which has demonstrated early promise (Tam et al., 2008).

In clinical studies involving small cohorts of hospitalized patients, both the Typhidot and TUBEX tests have demonstrated good performance in clinically suspected typhoid fever cases in comparison to the Widal test, particularly in early infection (Bhutta & Mansurali, 1999; House et al., 2001; Jesudason et al., 2002; Olsen et al., 2004; Begum et al., 2009; Narayanappa et al., 2010). In larger studies both in Asia and Africa, the new generation serological tests have compared less favourably (Dutta et al., 2006; Ley et al., 2011). Data from a large community-based surveillance study in Calcutta from 6697 patients with fever for 3 or more days demonstrated that, using a cut-off of fever for >5 days, the Widal test was more sensitive overall than the other two tests (Widal sensitivity 67%, specificity 85%, PPV 75%, NPV 79%; Typhidot 59%, 75%, 89% and 33%; Tubex 55%, 81%, 72% and 66%)(Dutta et al., 2006). The Widal test was also significantly cheaper but took longer to produce a result. One concern raised by the authors was that there was relatively poor standardisation of the kit reagents in the two newer tests and this may have had an effect due to the large number of tests performed.

More recently, the Dri-Dot Latex agglutination and IgM lateral flow assays have been developed by KIT Biomedical Research, Royal Tropical Institute, The Netherlands, and are simple to use for diagnosis of enteric fever. The validation study of the Dri-Dot Latex agglutination and IgM lateral flow assays for the diagnosis of typhoid fever, carried out in patients with clinically diagnosed typhoid fever in an Egyptian population, has demonstrated that the sensitivity and specificity were 71.4% and 86.3% for the Dri-Dot, and 80% and 71.4% for IgM Lateral Flow assay, respectively. A major limitation of these serologic tests is the limited sensitivity at the early stage of the disease. The sensitivity of these assays was increased to 84.3% when both tests were performed in parallel but the specificity decreased to 70.5%. Given that these assays are rapid and provide easy-to-interpret results, they may be useful for diagnosis of enteric fever in typhoid-endemic countries (Nakhla et al.,2011; Smith et al., 2011).

In summary, although several alternatives exist for diagnosing typhoid serologically, to-date the newer tests have not improved greatly on the performance of a test that is over a century old. With newer techniques for antigen discovery becoming available and an increasing amount of data being collected regarding the immune response to typhoid and paratyphoid infection, rapid and more effective diagnostic serological tests for typhoid infection are likely to become available in the near future.

3.3 Molecular diagnosis of enteric fever

Detecting the presence of S. Typhi in clinical samples using highly sensitive molecular techniques is not a recent development. In the 1980s, Rubin et al. designed and used a DNA probe cloned from *Citrobacter freundii* which has similar Vi antigen to S. Typhi for detection of S. Typhi and demonstrated 99% specificity and sensitivity using lactose-negative colonies or previously identified bacteria from febrile patients in Peru and in Indonesia (Rubin et al., 1988). As a direct diagnostic method however, the DNA probe method cannot detect less than 500 bacteria per ml of blood; patients with typhoid generally have fewer than 15 S. Typhi bacteria per ml (Watson, 1955; Wain et al., 1998). The DNA probe method was refined in a further study (Rubin et al., 1989), in which blood samples (and other specimens including bone marrow aspirates) were taken from patients presenting with febrile symptoms and concentrated by centrifugation using a DuPont Isolator tube, followed by overnight incubation of the bacteria on nylon filters. This modification allowed the detection of S. Typhi in 42% (13/32) of samples from patients with culture-confirmed typhoid fever using the equivalent of 2.5 ml of blood, compared with 53% (17/33) of these patients by culture of 8 ml peripheral blood. Additionally the probe detected 4 of 47 patients from whom S. Typhi was not isolated by culture, suggesting superior sensitivity could be achieved.

These early studies supported the introduction and development of further nucleic acid amplification tests to enable the rapid detection of very small numbers of bacterial components, thus providing new tools for sensitive and specific detection, identification and subsequently resistance testing of microorganisms starting from non-cultured sample material. Aside from the significant time saving over standard culture methods and the ability to detect much smaller number of bacteria, as with other organisms, nucleic acid amplification overcomes the issue of non-culturable or dead material, as is often seen with previous antibiotic treatment (Darton et al., 2009; Ho et al., 2009; Rello et al., 2009). After the early studies using DNA probes and hybridization techniques attention was turned to the use of polymerase chain reaction (PCR) methods for the detection of both S. Typhi and S. Paratyphi A for diagnosis of enteric fever.

3.3.1 Gene targets of PCR based assays for diagnosis of enteric fever

Generally any genomic sequences specific for S. Typhi or Paratyphi can be used as the PCR targets, and are easily available from the published DNA data bases. The widely researched targets for S. Typhi PCR-based assays include the S. Typhi flagellin gene *fliC-d* (Song et al., 1993; Hague et al., 1999, 2001; Kumar et al., 2002; Prakash et al., 2005; Ambati et al., 2007; Hatta & Smits 2007; Nandagopal et al., 2010; Nath et al., 2010), the *viaB* region encoding the Vi antigen of S. Typhi (Hashimoto et al., 1995), the *Salmonella* invasion gene *invA* (Cocolin et al., 1998), *hilA* gene encoding a transcription factor of S. Typhi (Sánchez-Jiménez & Cardona-Castro, 2004), Vi polysaccharide export ATP-binding protein *vexC* gene (Farrell et al., 2005), ST5 gene (Aziah et al., 2007), an iron-regulated gene *iroB* (Bäumler et al., 1997), 5S-23S spacer region (Zhu et al., 1996), and a heat shock protein *groEL* gene (Nair et al., 2002).

Other gene targets are also used in multiplex PCR assays, including the tyvelose epimerase gene (*tyv*; previously *rfbE*), *fliC-d*, *fliC-a* and the paratose synthase gene (*prt*; previously *rfbS*) (Hirose et al., 2002; Ali et al., 2009), *invA*, *viaB*, *fliC-d* and *prt* (Kumar et al., 2006), the outer membrane protein C (*ompC*), the putative regulatory protein gene STY4220, the intergenic region (SSPAI) between SSPA1723a and SSPA1724 in serovar Paratyphi A, and *stgA* (a fimbrial subunit protein) in serovar Typhi (Ngan et al., 2010), *stkF* (a putative fimbrial protein), *spa2473*, *spa2539*, *hsdM* (DNA methyltransferase) of S. Paratyphi (Ou et al., 2007).

Both S. Typhi and S. Paratyphi A have extremely limited genetic diversity within their populations and between 1 and 3% of the gene content of the S. Typhi and S. Paratyphi A genomes are unique (Roumagnac et al., 2006). This may aid DNA test specificity over other Gram-negative organisms. Further genomic exploration of both S. Typhi and S. Paratyphi A will identify new and better targets and then lead to novel nucleic acid based tests.

3.3.2 Sensitivity and specificity of PCR based assays for diagnosis of enteric fever

PCR-based tests for detecting the causative pathogens of enteric fever have developed rapidly over the last decade; however questions regarding the clinical utility and standardization of tests remain. Key to these issues is the array of methodologies used and variable sensitivities and specificities found. Song et al. (1993) was the first to apply PCR for detection of S. Typhi in clinical samples in an attempt to overcome the need for a pre-incubation or concentration step. Two pairs of oligonucleotide primers were designed to amplify the Hd flagella gene (*fliC-d*) of S. Typhi by nested PCR. This nested PCR had a minimum detection limit of 10 bacteria as determined by dilutions of DNA from S. Typhi and proved highly sensitive and specific using both laboratory and clinical samples. S. Typhi DNA was detected in 11 of 12 clinical specimens from patients with confirmed typhoid fever, whereas 10 blood specimens from patients with other febrile disease were all negative. Furthermore, this nested PCR also detected S. Typhi DNA from blood samples of 4 patients with suspected typhoid fever on the basis of clinical features but with negative cultures. Since then, many studies on the use of the nested PCR for detection of S. Typhi and diagnosis of typhoid fever have been published (Hague et al., 1999, 2001; Kumar et al., 2002; Prakash et al. 2005; Ambati et al., 2007; Hatta & Smits, 2007; Nath et al., 2010; Nandagopal et al., 2010). A nested PCR method was also developed using the *viaB* gene target, but its use in clinical diagnosis of enteric fever remains to be tested even though it demonstrated good sensitivity and specificity in tests performed on DNA samples isolated from clinical bacterial isolates (Hashimoto et al., 1995).

The nested PCR approach significantly improved the detection rate compared to that of blood culture and the Widal test; however its limitations include the longer time taken to perform and the more contaminations in comparison to a conventional PCR assay. Massi et al. utilized just one pair of primers ST1 and ST4 that Song et al. (1993) used for PCR detection of S. Typhi, and demonstrated that this single round PCR was also specific and could detect as little as 2-3 copies of S. Typhi DNA as determined by serial dilution of genomic DNA from S. Typhi (Massi et al., 2003). Using this conventional PCR method, genomic S. Typhi DNA was detected in 46 of 73 blood samples collected from patients with clinically suspected typhoid fever who had fever within 3 days of hospitalized admission, and who received no prior antibiotic treatment. PCR compared favourably (63% positivity amongst the clinically suspected cases) to blood culture (13.7%) and the Widal test (35.6%), using these 73 samples. The time taken for PCR analysis of each sample was less than 12 h, rather than 16 h for the nested PCR (Song et al., 1993) and between 3 to 5 days for blood culture.

Conventional PCR generally detects amplification using an agarose gel, which has limitations in sensitivity and speed. Cocolin et al. developed a PCR-microtitre plate hybridization technique for detection of S. Typhi *invA* by PCR, and demonstrated enhanced sensitivity and faster availability of results in comparison to a standard agarose gel electrophoresis approach (Cocolin et al., 1998). Other PCR assays were also researched on different gene targets in order to find a rapid and sensitive detection of S. Typhi in clinical specimens (Zhu et al., 1996; Bäumler et al., 1997; Nair et al., 2002; Sánchez-Jiménez & Cardona-Castro, 2004; Farrell et al., 2005; Nizami et al., 2006; Aziah et al., 2007).

Real-time PCR (RT-PCR), which is generally detected by measuring a fluorescent signal and has several advantages over conventional PCR has recently been explored, yet not exhaustively, for detection of both S. Typhi and S. Paratyphi A. Massi et al. applied TaqMan-based real-time PCR (TaqMan assay) to the quantification of S. Typhi in the blood of patients suspected of having typhoid fever by targeting the S. Typhi flagellin gene in genomic DNAs isolated from blood samples (Massi et al., 2005). Of 55 blood samples taken from suspected typhoid fever patients, eight blood samples with a positive blood culture had S. Typhi loads ranging from 1.01×10^3 to 4.35×10^4 copies/ml blood, and from 47 blood samples with negative blood culture, there were 40 (85.1%) TaqMan assay-positive samples with loads ranging from 3.9 to 9.9×10^2 copies/ml blood. In their study, the TaqMan assay detected more than 10^3 copies/ml blood of S. Typhi in all of the blood culture-positive samples, whereas less than 10^3 copies/ml blood of S. Typhi were detected in the blood culture-negative samples. This suggests that a TaqMan assay may be useful for assessing S. Typhi loads, especially in cases of suspected typhoid fever with negative results from the standard blood culture test.

Farrell et al. developed broad-range (Pan) *Salmonella* and *S.* Typhi specific real-time PCR assays using LightCycler (Roche Diagnostics, Indianapolis, IN). Using direct stool samples the pan-*Salmonella* assay was validated with 96% (53/55) sensitivity and 96% (49/51) specificity. However, the *S.* Typhi-specific PCR assay was not sufficiently validated due to the low incidence of *S.* Typhi infections in the test region (Farrel et al., 2005).

All these studies demonstrated that the sensitivity and specificity of PCR assays was significantly better compared to that of blood culture and/or the Widal test, and some selected evaluation studies of these tests are summarized in Table 1.

Test used	Target gene	Samples (n) tested		Blood culture	PCR	Widal test	Reference
nested PCR	*fliC-d*	suspected	16	12BC+ 4BC-	11/12BC+ 4/4BC-		Song et al. 1993
		control	10 febrile		0/10		
nested PCR	*fliC-d*	suspected	55	8BC+ 47BC-	8/8BC+ 24/47BC-	6/8BC+ 23/47BC-	Hague et al. 2001
		control	20 nonfebrile		0/20	9/20	
nested PCR	*fliC-d*	suspected	40	20BC+ 20BC-	20/20BC+ 12/20BC-		Kumar et al. 2002
		control	None				
nested PCR	*fliC-d*	suspected	63	17BC+ 46BC-	17/17BC+ 36/46BC-	12/17BC+ 4/46BC-	Prakash et al. 2005
		Control	25 nonfebrile		0/25	1/25	
nested PCR	*fliC-d*	suspected	119	68BC+ 51BC-	67/68BC+ 26/51BC-	34/68BC+ 11/51BC-	Hatta & Smiths 2007
		control	12 febrile		0/12	4/12	
nested PCR	*fliC-d*	suspected	42	14BC+ 38BC-	14/14BC+ 29/38BC-	7/14BC+ 19/38BC-	Ambati et al. 2007
		control	11 febrile 8 nonfebrile		0/11 0/8	2/11 0/8	
nested PCR	*fliC-d*	suspected	291	6BC+ 285BC-	6/6BC+ 8/285BC-		Nandagopal et al. 2010
		control	10 febrile		0/10		
PCR	*viaB*	suspected	203	26 BC+ 177BC-	10/26BC+ 12/177BC-		Nizami et al. 2006
		control	None				
PCR	*hilA*	suspected	37	34BC+ 3BC-	34/34BC+ 3/3BC-		Sánchez-Jiménez & Cardona-Castro 2004
		control	35 infected with other pathogens		0/35		
			150 healthy volunteers		0/150		

Test used	Target gene	Samples (n) tested		Blood culture	PCR	Widal test	Reference
PCR	*fliC-d*	suspected	82	28BC+	59/82		Haque et al. 1999
		control	20 nonfebrile		0/20		
PCR	*fliC-d*	suspected	73	10BC+ 63BC-	10/10BC+ 36/63BC-	10/10BC+ 16/63BC-	Massi et al. 2003
		Control	None				
PCR	ST-50	suspected	33BC+ broths		29/33		Aziah et al. 2007
		control	40BC- broths		0/40		
PCR	*fliC-d*	suspected	820	78BC+ 742BC-	73/78BC+ 95/742BC-		Chaudhry et al. 2010
		control	None				
RT-PCR	*fliC-d*	suspected	55	8BC+ 47BC-	8/8BC+ 40/47BC-		Massi et al. 2005
		control	26 nonfebrile		0/26		

BC: Blood culture; BC+: Blood culture positive; BC-: Blood culture negative

Table 1. The results of selected studies on the sensitivity and specificity of PCR, blood culture and Widal test on blood samples from patients with suspected enteric fever

3.3.3 Multiplex PCR detection for *S.* Typhi and *S.* Paratyphi

Classically *S.* Typhi has been considered as the major cause of enteric fever; however, in recent years *S.* Paratyphi and Vi-negative variants of *S.* Typhi have emerged rapidly (Wain et al., 2005; Dong et al., 2010). *S.* Paratyphi A is a causative agent of paratyphoid fever and has become a major cause of enteric fever in Asia. For example, more than 80% of enteric fever outbreaks have been caused by *S.* Paratyphi since 1998, three years after Vi polysaccharide typhoid fever vaccine was introduced in Guangxi province China (Dong et al., 2010). The largest one (495 episodes), which occurred in 2004 in Luocheng County, was caused by a contaminated water supply system. *S.* Paratyphi has been the predominant cause of enteric fever in Guangxi province China since 1999 (Dong et al., 2010). Studies from India and Nepal also suggested that paratyphoid fever caused by *S.* Paratyphi A can contribute up to half of all cases of enteric fever in some settings (Ochiai et al., 2005; Woods et al., 2006). PCR tests using *S.* Typhi specific primers appear to be sensitive to detect typhoid fever, but cannot detect paratyphoid fever. Recent developments in multiplex PCR methods have addressed the issue of paratyphoid as well as typhoid fever diagnosis.

Hirose et al. developed a complex PCR using the primers for O, H, and Vi antigen genes, *tyv* (*rfbE*), *prt* (*rfbS*), *fliC-d*, *fliC-a*, and *viaB*, for the rapid identification of *S.* Typhi and *S.* Paratyphi A. This assay was able to accurately identify and distinguish *S.* Typhi and *S.* Paratyphi A from laboratory isolates; however, its clinical use was not assessed (Hirose et al., 2002). Similarly, Levy et al. developed a multiplex PCR to identify *Salmonella* serogroups A, B and D, and Vi-positive strains. Blinded testing of 664 Malian and Chilean *Salmonella* blood isolates demonstrated 100% sensitivity and specificity; again clinical utility was not assessed (Levy et al. 2008). Kumar et al. explored another set of target genes including those responsible for invasion (*invA*), O (*prt*), H (*fliC-d*) and Vi (*viaB*) antigen genes in a multiplex PCR, and demonstrated accurate identification of laboratory isolates and 100% detection probability when a cell suspension of 10^4 CFU/ml (500 CFU per reaction) was used. *S*. Typhi bacteria were artificially inoculated into water and food (milk and meat rinse) samples and detected by the multiplex PCR after overnight pre-enrichment in buffered peptone water. No *Salmonella* bacteria could be detected from water samples collected from the field by the multiplex PCR or standard culture method (Kumar et al., 2006).

Using the same target genes as Hirose et al. (Hirose et al., 2002), Ali et al. further optimised the primers and applied the nested multiplex PCR directly to clinical blood specimens for diagnosis. Of 42 multiplex PCR-positive blood samples, they showed that 26, 9, and 2 were Vi-positive *S*. Typhi, Vi-negative *S*. Typhi and *S*. Paratyphi A, respectively, and five patients had a mixed infection. Tests with several common pathogens confirmed that the assay was specific (Ali et al., 2009).

The analysis of the genome of *S*. Paratyphi led Ou et al. to identify four gene targets (*stkF*, *spa2473*, *spa2539* and *hsdM*) which were used to develop a highly discriminatory multiplex PCR assay (Ou et al., 2007). A valuation study using spiked blood and stool samples demonstrated that the sensitivity of the discriminatory multiplex PCR was 1×10^5 CFU/ml and 2×10^5 CFU/ml, respectively, and however, the sensitivity can be increased to 1×10^4 CFU/ml and 2×10^3 CFU/ml after 5 h culture enrichment (Teh et al., 2008). Nagarajan et al. have further improved upon the existing PCR-based diagnostic technique by using one pair of primers that is unique to *S*. Typhi and *S*. Paratyphi A, corresponding to the STY0312 gene in *S*. Typhi and its homolog SPA2476 in *S*. Paratyphi A, and another pair that amplifies the region in *S*. Typhi CT18 and *S*. Typhi Ty2 corresponding to the region between genes STY0313 to STY0316 but which is absent in *S*. Paratyphi A. The possibility of a false-negative result arising due to mutation in hypervariable genes has been reduced by targeting a gene unique to typhoidal *Salmonella* serovars as a diagnostic marker. This set of primers can also differentiate between *S*. Typhi CT18, *S*. Typhi Ty2, and *S*. Paratyphi A, which have stable deletions in this specific locus. The PCR assay designed in this study has a sensitivity of 95% compared to the Widal test which has a sensitivity of only 63% (Nagarajan et al., 2009). Ngan et al. developed another multiplex PCR format in which the outer membrane protein C (*ompC*) was used for detection of members of the *Salmonella* genus, the putative regulatory protein gene STY4220 for the presence of either *S*. Typhi or *S*. Paratyphi A, and the intergenic region (SSPAI) between SSPA1723a and SSPA1724 in serovar Paratyphi A and a fimbrial subunit protein (*stgA*) in serovar Typhi for differentiation between *S*. Typhi and *S*. Paratyphi. This multiplex PCR was evaluated using 124 clinical and reference *Salmonella* serovars and both *S*. Typhi and *S*. Paratyphi A were detected at 100% specificity and sensitivity. This multiplex PCR reaction can detect approximately 1 pg of *Salmonella* genomic DNA. When tested on 8 h enriched spiked blood samples of serovars Typhi and Paratyphi A, the sensitivity was estimated at 4.5×10^4 - 5.5×10^4 CFU/ml, with similar detection levels observed for spiked fecal samples (Ngan et al., 2010).

Recently Nga et al. used a novel multiplex three colour real-time PCR assay to detect specific target sequences in the genomes of *S*. Typhi and *S*. Paratyphi A. The assay was validated and demonstrated a high level of specificity and reproducibility under experimental conditions with the DNA extracted from blood and bone marrow samples from culture positive and negative enteric fever patients. All bone marrow samples tested were positive for *Salmonella*; however, the sensitivity on blood samples was limited. The assay demonstrated an overall specificity of 100% (75/75) and sensitivity of 53.9% (69/128) on biological samples. The data on the PCR detection limit suggested that PCR performed directly on blood samples may be an unsuitable methodology and a potentially unachievable target for the routine diagnosis of enteric fever because the bacterial load of *S.* Typhi in peripheral blood is low, often below the limit of detection by culture and, consequently, below detection by PCR (Nga et al., 2010).

3.3.4 Novel blood culture PCR system and application in human challenge study

An alternative strategy to increase the sensitivity and specificity of PCR is PCR amplification on the blood culture after a short period of incubation. We have recently developed a fast and highly sensitive blood culture PCR method for detection of *Salmonella* serovar Typhi (Zhou & Pollard, 2010). The method uses an optimised ox bile tryptone soy broth for blood culture with subsequent PCR assay in an attempt to reduce the turn-around time for diagnosis and increase diagnostic sensitivity. By using a 5-hour incubation, 3 CFU of *S.* Typhi cells could mutliply over about 10 generations. This was assessed by a time-course experiment, the results of which were published (Zhou & Pollard, 2010) and are cited here in Table 2.

Incubation time (hour)	CFU[a]	*fliC-d* amplicons[b]
0	3	---
1	4	---
2	17	----
3	105	+++
4	209	+++
5	4461	+++

*Three bacteria of *Salmonella* serovar Typhi were incubated in the tryptone soy broth containing 2.4% ox bile and 20% blood. [a] The mean of three independent experiments; [b] *Salmonella* serovar Typhi *fliC-d* amplicons resulting from PCR using the DNA templates prepared from three independent cultures.

Table 2. The growth and PCR detection of *S.* Typhi in ox bile tryptone soy broth blood culture*.

The sensitivity of this blood culture-PCR method was equivalent to 0.75 CFU per millilitre of blood which is similar to the level of clinical typhoid samples which regular PCR cannot detect. The whole blood culture PCR assay takes less than 8 hours to complete rather than several days for conventional blood culture. This novel blood culture PCR method is superior in speed and sensitivity to both conventional blood culture and PCR assays. Its use in clinical diagnosis may allow early detection of the causative organism and facilitate initiation of prompt treatment among patients with typhoid fever. The recent use of this novel culture PCR method to our ongoing human typhoid challenge studies has proved that the advantage of combining culture and PCR amplification is an increase in the speed of a positive confirmatory diagnosis, even though it is unlikely to produce a greater level of sensitivity than that of traditional culture alone. However, practical clinical use in diagnosing enteric fever of this culture PCR system remains to be proved, in particular, using blood samples with antibiotic pre-treatment.

4. Future perspectives

Blood culture has some distinct advantage over other diagnostic methods, such as the combination of bacterial identification with antibiotic susceptibility, and an unquestioned role in providing epidemiological data; however, it has many problems related to its relatively long turnaround time and low sensitivity, especially in patients receiving antibiotic treatment. Detection of bacterial DNA in whole blood by PCR assay is the methodology most able to substantially decrease the turnaround time without bias from the inhibitory effect of antibiotics, yet the published PCR assays for diagnosis of enteric fever are in limited use. Further investigation to develop rapid and reliable diagnostics for enteric fever are urgently needed.

One of the limiting factors in the use of current PCR methodology in clinical diagnosis of enteric fever is the low number of bacteria circulating in the blood of enteric fever patients. Advancement in the use of PCR would require the capture and amplification of a smaller number of bacteria (maybe even a single organism) in blood or other bodily fluids. Such a task is not insurmountable but it will be a challenge to make it cost effective (Baker et al., 2010). An alternative approach to increase the PCR assay sensitivity and specificity is to remove the interfering human genomic DNA present in the samples. To achieve this, selective lysis of human genomic DNA with external nuclease may be usful, as proven in pathogen identification in patients with sepsis (Horz et al., 2008; Handschur et al., 2009). Removal of dominant human genomic DNA causes enrichment of bacterial DNA, thus improving sensitivity and specificity of PCR assays. Using S. Typhi spiked blood samples, we have demonstrated that this approach can increase the sensitivity of PCR assays by more than 1,000 fold (unpublished result). However, a field trial with clinical typhoid specimens is needed to confirm the laboratory findings. Reverse transcription PCR may be another choice to detect such a low number of bacteria in typhoid patient blood as the higher number of copies of mRNA for a specific gene target could increase the PCR assay sensitivity. The *fliC* of S. Typhi was used as target in the reverse transcription-multiplex PCR assay for simultaneous detection of *Escherichia coli* O157:H7, *Vibrio cholerae* O1 and S. Typhi (Morin et al., 2004).

The study on host specific responses to enteric fever may identify signatures of host-pathogen interactions with S. Typhi, which will form the basis of development of new molecular diagnostics for enteric fever. Activation of host specific genes or pathways during infection could be identified using DNA microarrays; a physiological signature or metabolic product associated with typhoid could be studied with mass spectrometry or other proteomic technologies. For example, surface-enhanced laser desorption/ionization time-of-flight (SELDI-TOF) mass spectroscopy has been used in identifying SARS protein biomarkers (Mazzulli et al., 2005). All these new technological approaches may add insight into proteins as biomarkers of typhoid infection, and potentially result in a new generation of novel molecular diagnostics for enteric fever.

Enteric fever is endemic in resource poor countries, and development in new technologies should focus on how these can be applied to location with limited resources. Efforts are being made to simplify typhoid PCR assays using pre-prepared and freeze-dried regents (Aziah et al., 2007). However, new PCR technologies, such as isothermal PCR, are of particularly practical use in the diagnosis of enteric fever, as these methods allow for the possibility of developing less-complicated and less-expensive machinery than is necessary for conventional PCR. Several isothermal PCR technologies have been developed (Gill & Ghaemi, 2008), including strand displacement amplification (SDA) (Walker et al., 1992), loop-mediated amplification (LAMP) (Notomi et al., 2000), and helicase-dependent amplification (HDA) (Vincent et al., 2004). Recently, Francois et al. have examined the robustness of LAMP for bacterial diagnostic applications using *S.* Typhi as the target organism (Francois et al., 2011), and demonstrated that LAMP is more sensitive than conventional qPCR and is also a very robust, innovative and powerful molecular diagnostic method. However, SDA, HDA and/or other isothermal amplification methods could be more advantageous over LAMP in multiplex amplifications. The recent surge in paratyphoid disease makes it necessary to develop new diagnostics for detection of both S. Typhi and Paratyphi. Another advantage of isothermal PCR is its potential for use in resource poor or point-of-care settings.

In summary, advancement in genomics and proteomics will further our understanding of molecular pathogenesis of enteric fever, and eventually lead to identification of new targets which could form the basis for new molecular diagnostics. With progress in new technologies, we expect that a new generation of fast and sensitive molecular diagnostics for enteric fever will be developed in the near future.

5. Acknowledgments

LZ and TD are funded by a grant from the Wellcome Trust, UK and CW is funded by the UK National Institute for Health Research (NIHR) Oxford Biomedical Research Centre.

6. References

Ackers, M.L.; Puhr, N.D.; Tauxe, R.V. & Mintz, E.D. (2000). Laboratory-based surveillance of *Salmonella* serotype Typhi infections in the United States: antimicrobial resistance on the rise. JAMA. 2000 May 24-31;283(20):2668-73.

Albaqali, A.; Ghuloom, A.; Al Arrayed, A.; Al Ajami, A.; Shome, D.K.; Jamsheer, A.; Al Mahroos, H.; Jelacic, S.; Tarr, P.I.; Kaplan, B.S. & Dhiman, R.K. (2003). Hemolytic uremic syndrome in association with typhoid fever. Am J Kidney Dis. 2003 Mar; 41(3):709-13.

Ali, A.; Haque, A.; Haque, A.; Sarwar, Y.; Mohsin, M.; Bashir, S. & Tariq, A. (2009). Multiplex PCR for differential diagnosis of emerging typhoidal pathogens directly from blood samples. Epidemiol Infect. 2009 Jan;137(1):102-7.

Al Naiemi, N.; Zwart, B.; Rijnsburger, M.C.; Roosendaal, R.; Debets-Ossenkopp, Y.J.; Mulder, J.A.; Fijen, C.A.; Maten, W.; Vandenbroucke-Grauls, C.M. & Savelkoul, P.H. (2008). Extended-spectrum-beta-lactamase production in a *Salmonella enterica* serotype Typhi strain from the Philippines. J Clin Microbiol. 2008 Aug;46(8):2794-5.

Ambati, S.R.; Nath, G. & Das, B.K. (2007). Diagnosis of typhoid fever by polymerase chain reaction. Indian J Pediatr. 2007 Oct;74(10):909-13.

Aziah, I.; Ravichandran, M. & Ismail, A. (2007). Amplification of ST50 gene using dry-reagent-based polymerase chain reaction for the detection of *Salmonella* Typhi. Diagn Microbiol Infect Dis. 2007 Dec;59(4):373-7.

Baker, S.; Favorov, M. & Dougan, G. (2010). Searching for the elusive typhoid diagnostic. BMC Infect Dis. 2010 Mar 5;10:45.

Balasubramanian, S.; Shivbalan, S. & Miranda, P.K. (2003). Pseudotumour cerebri as an unusual manifestation of typhoid. Ann Trop Paediatr. 2003; 23: 223–24.

Bakr, W.M.; El Attar, L.A.; Ashour, M.S. & El Toukhy, A.M. (2011). The dilemma of widal test - which brand to use? a study of four different widal brands: a cross sectional comparative study. Ann Clin Microbiol Antimicrob. 2011 Feb 8;10:7.

Bäumler, A.J.; Heffron, F. & Reissbrodt, R. (1997). Rapid detection of *Salmonella enterica* with primers specific for iroB. J Clin Microbiol. 1997 May;35(5):1224-30.

Begum, Z.; Hossain, M.A.; Musa, A.K.; Shamsuzzaman, A.K.; Mahmud, M.C.; Ahsan, M.M.; Sumona, A.A.; Ahmed, S.; Jahan, N.A.; Alam, M. & Begum, A. (2009). Comparison between DOT EIA IgM and Widal Test as early diagnosis of typhoid fever. Mymensingh Med J. 2009 Jan;18(1):13-7.

Bhan, M.K.; Bahl, R. & Bhatnagar, S. (2005). Typhoid and paratyphoid fever. Lancet. 2005 Aug 27-Sep 2;366(9487):749-62.

Bhan, M.K.; Bahl, R.; Sazawal, S.; Sinha, A.; Kumar, R.; Mahalanabis, D. & Clemens, J.D. (2002). Association between *Helicobacter pylori* infection and increased risk of typhoid fever. J Infect Dis. 2002 Dec 15;186(12):1857-60.

Bhutta, Z. A. (2006). Current concepts in the diagnosis and treatment of typhoid fever. BMJ. 2006 Jul 8;333(7558):78-82.

Bhutta, Z.A. & Mansurali, N. (1999). Rapid serologic diagnosis of pediatric typhoid fever in an endemic area: a prospective comparative evaluation of two dot-enzyme immunoassays and the Widal test. Am J Trop Med Hyg. 1999 Oct;61(4):654-7.

Butler, T.; Islam, A.; Kabir, I. & Jones, P.K. (1991). Patterns of morbidity and mortality in typhoid fever dependent on age and gender: review of 552 hospitalized patients with diarrhea. Rev Infect Dis. 1991 Jan-Feb;13(1):85-90.

Chaudhry, R.; Chandel, D.S.; Verma, N.; Singh, N.; Singh, P. & Dey, A.B. (2010). Rapid diagnosis of typhoid fever by an in-house flagellin PCR. J Med Microbiol. 2010 Nov;59(Pt 11):1391-3.

Chau, T.T.; Campbell, J.I.; Galindo, C.M.; Van Minh Hoang, N.; Diep, T.S.; Nga, T.T.; Van Vinh Chau, N.; Tuan, P.Q.; Page, A.L.; Ochiai, R.L.; Schultsz, C.; Wain, J.; Bhutta, Z.A.; Parry, C.M.; Bhattacharya, S.K.; Dutta, S.; Agtini, M.; Dong, B.; Honghui, Y.; Anh, D.D.; Canh do, G.; Naheed, A.; Albert, M.J.; Phetsouvanh, R.; Newton, P.N.; Basnyat, B.; Arjyal, A.; La, T.T.; Rang, N.N.; Phuong le, T.; Van Be Bay, P.; von Seidlein, L.; Dougan, G.; Clemens, J.D.; Vinh, H.; Hien, T.T.; Chinh, N.T.; Acosta, C.J.; Farrar, J. & Dolecek, C. (2007). Antimicrobial drug resistance of *Salmonella enterica* serovar Typhi in asia and molecular mechanism of reduced susceptibility to the fluoroquinolones. Antimicrob Agents Chemother. 2007 Dec;51(12):4315-23.

Chew, S.K.; Cruz, M.S. & Lim, Y.S. (1992). Monteiro EH. Diagnostic value of the Widal test for typhoid fever in Singapore. J Trop Med Hyg. 1992 Aug;95(4):288-91.

Chiu, C.H. & Lin, T.Y. (1999). Typhoid fever in children. Lancet. 1999 Dec 4;354(9194):2001-2.

Choo, K.E.; Davis, T.M.; Ismail, A.; Tuan Ibrahim, T.A. & Ghazali, W.N. (1999). Rapid and reliable serological diagnosis of enteric fever: comparative sensitivity and specificity of Typhidot and Typhidot-M tests in febrile Malaysian children. Acta Trop. 1999 Mar 15;72(2):175-83.

Choo, K.E.; Oppenheimer, S.J.; Ismail, A.B. & Ong, K.H. (1994). Rapid serodiagnosisof typhoid feverby dot enzyme immunoassay in an endemic area. Clin Infect Dis. 1994 Jul;19(1):172-6.

Clark, T.W.; Daneshvar, C.; Pareek, M.; Perera, N. & Stephenson, I. (2010). Enteric fever in a UK regional infectious diseases unit: a 10 year retrospective review. J Infect. 2010 Feb;60(2):91-8.

Cocolin, L.; Manzano, M.; Astori, G.; Botta, G.A; Cantoni, C. & Comi, G. (1998). A highly sensitive and fast non-radioactive method for the detection of polymerase chain reaction products from *Salmonella* serovars, such as *Salmonella* Typhi, in blood specimens. FEMS Immunol Med Microbiol. 1998 Nov;22(3):233-9.

Coleman, W. &. Buxton, B.H. (1907). The bacteriology of the blood in typhoid fever. Amer J Med Sci. 1907;133:896-903.

Crawford, R.W.; Rosales-Reyes, R.; Ramírez-Aguilar Mde, L.; Chapa-Azuela, O.; Alpuche-Aranda, C. & Gunn, J.S. (2010). Gallstones play a significant role in *Salmonella* spp. gallbladder colonization and carriage. Proc Natl Acad Sci U S A. 2010 Mar 2;107(9):4353-8.

Crump, J.A.; Luby, S.P. & Mintz, E.D. (2004). The global burden of typhoid fever. Bull World Health Organ. 2004 May;82(5):346-53.

Darton, T.; Guiver, M.; Naylor, S.; Jack, D.L.; Kaczmarski, E.B.; Borrow, R. & Read, R.C. (2009). Severity of meningococcal disease associated with genomic bacterial load. Clin Infect Dis. 2009 Mar 1;48(5):587-94.

Datta, V.; Sahare, P. & Chaturved, P. (2004). Guillain-Barre syndrome as a complication of enteric fever. J Indian Med Assoc. 2004 Mar;102(3):172-3.

Dong, B.Q.; Yang, J.; Wang, X.Y.; Gong, J.; von Seidlein, L.; Wang, M.L.; Lin, M.; Liao, H.Z.; Ochiai, R.L.; Xu, Z.Y.; Jodar, L. & Clemens, J.D. (2010). Trends and disease burden of enteric fever in Guangxi province, China, 1994-2004. Bull World Health Organ. 2010 Sep 1;88(9):689-96.

Dutta, S.; Sur, D.; Manna, B.; Sen, B.; Deb, A.K.; Deen, J.L.; Wain, J.; Von Seidlein, L.; Ochiai, L.; Clemens, J.D. & Kumar, B.hattacharya, S. (2006). Evaluation of new-generation serologic tests for the diagnosis of typhoid fever: data from a community-based surveillance in Calcutta, India. Diagn Microbiol Infect Dis. 2006 Dec;56(4):359-65.

Meltzer, E.; Sadik, C. & Schwartz, E. (2005). Enteric fever in Israeli travelers: a nationwide study. J Travel Med. 2005 Sep-Oct;12(5):275-81.

Maskey, A.P.; Day, J.N.; Phung, Q.T.; Thwaites, G.E.; Campbell, J.I.; Zimmerman, M.; Farrar, J.J. & Basnyat, B. (2006). *Salmonella enterica* serovar Paratyphi A and S. enterica serovar Typhi cause indistinguishable clinical syndromes in Kathmandu, Nepal. Clin Infect Dis. 2006 May 1;42(9):1247-53.

Ekdahl, K.; De Jong, B. & Andersson, Y. (2005). Risk of travel-associated typhoid and paratyphoid fevers in various regions. J Travel Med. 2005 Jul-Aug;12(4):197-204.

Fadeel, M.A.; House, B.L.; Wasfy, M.M.; Klena, J.D.; Habashy, E.E.; Said, M.M.; Maksoud, M.A.; Rahman, B.A. & Pimentel, G. (2011). Evaluation of a newly developed ELISA against Widal, TUBEX-TF and Typhidot for typhoid fever surveillance. J Infect Dev Ctries. 2011 Mar 21;5(3):169-75.

Fangtham, M. & Wilde, H. (2008). Emergence of *Salmonella Paratyphi A* as a Major Cause of Enteric Fever: Need for Early Detection, Preventive Measures, and Effective Vaccines. J Travel Med. 2008 Sep-Oct;15(5):344-50.

Farrell, J.J.; Doyle, L.J.; Addison, R.M.; Reller, L.B.; Hall, G.S. & Procop, G.W. (2005). Broad-range (pan) *Salmonella* and *Salmonella* serotype Typhi-specific real-time PCR assays: potential tools for the clinical microbiologist. Am J Clin Pathol. 2005 Mar;123(3):339-45.

Feleszko, W.; Maksymiuk, J.; Oracz, G.; Golicka, D. & Szajewska, H. (2004). The TUBEX typhoid test detects current *Salmonella* infections. J Immunol Methods. 2004 Feb 1;285(1):137-8.

Francois, P.; Tangomo, M.; Hibbs, J.; Bonetti, E.J.; Boehme, C.C.; Notomi, T.; Perkins, M.D. & Schrenzel, J. (2011). Robustness of a loop-mediated isothermal amplification reaction for diagnostic applications. FEMS Immunol Med Microbiol. 2011 Jun;62(1):41-8.

Gasem, M.H.; Smits, H.L.; Goris, M.G. & Dolmans, W.M. (2002). Evaluation of a simple and rapid dipstick assay for the diagnosis of typhoid fever in Indonesia. J Med Microbiol. 2002 Feb;51(2):173-7.

Gill, P. & Ghaemi, A. (2008). Nucleic acid isothermal amplification technologies: a review. Nucleosides Nucleotides Nucleic Acids. 2008 Mar;27(3):224-43.

Gilman, R.H.; Terminel, M.; Levine, M.M.; Hernandez-Mendoza, P. & Hornick, R.B. (1975). Relative efficacy of blood, urine, rectal swab, bone-marrow, and rose-spot cultures for recovery of *Salmonella* Typhi in typhoid fever. Lancet. 1975 May 31;1(7918):1211-3.

Glynn, J.R.; Hornick, R.B.; Levine, M.M.& Bradley, D.J. (1995). Infecting dose and severity of typhoid: analysis of volunteer data and examination of the influence of the definition of illness used. Epidemiol Infect. 1995 Aug;115(1):23-30.

Gotuzzo, E.; Frisancho, O.; Sanchez, J.; Liendo, G.; Carrillo, C.; Black, R.E. & Morris, J.G. Jr. (1991). Association between the acquired immunodeficiency syndrome and infection with *Salmonella* Typhi or *Salmonella* paratyphi in an endemic typhoid area. Arch Intern Med. 1991 Feb;151(2):381-2.

Handschur, M.; Karlic, H.; Hertel, C.; Hertel, C.; Pfeilstöcker, M. & Haslberger, A.G. (2009). Preanalytic removal of human DNA eliminates false signals in general 16S rDNA PCR monitoring of bacterial pathogens in blood. Comp Immunol Microbiol Infect Dis. 2009 May;32(3):207-19.

Haque, A.; Ahmed, J.; & Qureshi, J.A. (1999). Early detection of typhoid by polymerase chain reaction. Ann Saudi Med. 1999 Jul-Aug;19(4):337-40.

Haque, A.; Ahmed, N.; Peerzada, A.; Raza, A.; Bashir, S. & Abbas, G. (2001). Utility of PCR in diagnosis of problematic cases of typhoid. Jpn J Infect Dis. 2001 Dec;54(6):237-9.

Hashimoto, Y.; Itho, Y.; Fujinaga, Y.; Khan, A.Q.; Sultana, F.; Miyake, M.; Hirose, K.; Yamamoto, H. & Ezaki, T. (1995). Development of nested PCR based on the ViaB sequence to detect *Salmonella* Typhi. J Clin Microbiol. 1995 Nov;33(11):3082.

Hatta, M. & Smits, H.L. (2007). Detection of *Salmonella* Typhi by nested polymerase chain reaction in blood, urine, and stool samples. Am J Trop Med Hyg. 2007 Jan;76(1):139-43.

Hatta, M.; Goris, M.G.; Heerkens, E.; Gooskens, J. & Smits, H.L. (2002). Simple dipstick assay for the detection of *Salmonella* Typhi-specific IgM antibodies and the evolution of the immune response in patients with typhoid fever. Am J Trop Med Hyg. 2002 Apr;66(4):416-21.

Hirose, K.; Itoh, K.; Nakajima, H.; Kurazono, T.; Yamaguchi, M.; Moriya, K.; Ezaki, T.; Kawamura, Y.; Tamura, K. & Watanabe, H. (2002). Selective amplification of tyv (rfbE), prt (rfbS), viaB, and fliC genes by multiplex PCR for identification of *Salmonella enterica* serovars Typhi and Paratyphi A. J Clin Microbiol. 2002 Feb;40(2):633-6.

Hoffman, S.L.; Edman, D.C.; Punjabi, N.H.; Lesmana, M.; Cholid, A., Sundah, S. & Harahap, J. (1986). Bone marrow aspirate culture superior to streptokinase clot culture and 8 ml 1:10 blood-to-broth ratio blood culture for diagnosis of typhoid fever. Am J Trop Med Hyg. 1986 Jul;35(4):836-9.

Horz, H.P.; Scheer, S.; Huenger, F.; Vianna, M.E. & Conrads, G. (2008). Selective isolation of bacterial DNA from human clinical specimens. J Microbiol Methods. 2008 Jan;72(1):98-102.

Ho, Y.C.; Chang, S.C.; Lin, S.R. & Wang, W.K. (2009). High levels of mecA DNA detected by a quantitative real-time PCR assay are associated with mortality in patients with methicillin-resistant *Staphylococcus aureus* bacteremia. J Clin Microbiol. 2009 May;47(5):1443-51.

House, D.; Wain, J.; Ho, V.A.; Diep, T.S.; Chinh, N.T., Bay, P.V.; Vinh, H.; Duc, M.; Parry, C.M.; Dougan, G.; White, N.J. Hien, T.T. & Farrar, J.J. (2001). Serology of typhoid fever in an area of endemicity and its relevance to diagnosis. J Clin Microbiol. 2001 Mar;39(3):1002-7.

Ismail, A.; Kader, Z.S. & Ong, K.H. (1991). Dot enzyme immunosorbent assay for the serodiagnosis of typhoid fever. Southeast Asian J Trop Med Public Health. 1991 Dec;22(4):563-6.

Jesudason, M.; Esther, E. & Mathai, E. (2002). Typhidot test to detect IgG & IgM antibodies in typhoid fever. Indian J Med Res. 2002 Aug;116:70-2.

Kawano, R.L.; Leano, S.A. & Agdamag, D.M. (2007). Comparison of serological test kits for diagnosis of typhoid fever in the Philippines. J Clin Microbiol. 2007 Jan;45(1):246-7.

Kaye, D.; Palmieri, M. & Rocha, H. (1966). Effect of bile on the action of blood against *Salmonella*. J Bacteriol. 1966 Mar;91(3):945-52.

Khosla, S.N.; Jain, N. & Khosla, A. (1993). Gastric acid secretion in typhoid fever. Postgrad Med J. 1993 Feb;69(808):121-3.

Kumar, A.; Arora, V.; Bashamboo, A. & Ali, S. (2002). Detection of *Salmonella* Typhi by polymerase chain reaction: implications in diagnosis of typhoid fever. Infect Genet Evol. 2002 Dec;2(2):107-10.

Kumar, S.; Balakrishna, K. & Batra, H.V. Detection of *Salmonella enterica* serovar Typhi (*S.* Typhi) by selective amplification of invA, viaB, fliC-d and prt genes by polymerase chain reaction in mutiplex format. Lett Appl Microbiol. 2006 Feb;42(2):149-54.

Kundu, R.; Ganguly, N., Ghosh, T.K.; Yewale, V.N.; Shah, R.C. & Shah, N.K. (2006). IAP Task Force. IAP Task Force Report: diagnosis of enteric fever in children. Indian Pediatr. 2006 Oct;43(10):875-83.

Lang, R.; Maayan, M.C.; Lidor, C.; Savin, H.; Kolman, S. & Lishner, M. (1992). *Salmonella* Paratyphi C osteomyelitis: report of two separate episodes 17 years apart. Scand J Infect Dis. 1992; 24: 793–96.

Lee, W.S.; Puthucheary, S.D. & Parasakthi, N. (2000). Extra-intestinal nontyphoidal *Salmonella* infections in children. Ann Trop Paediatr. 2000; 20: 125–29.

Levine, M.M.; Black, R.E. & Lanata, C. (1982). Precise estimation of the numbers of chronic carriers of *Salmonella* Typhi in Santiago, Chile, an endemic area. J Infect Dis. 1982 Dec;146(6):724-6.

Levine, M.M.; Grados, O.; Gilman, R.H.; Woodward, W.E.; Solis-Plaza, R. & Waldman, W. (1978). Diagnostic value of the Widal test in areas endemic for typhoid fever. Am J Trop Med Hyg. 1978 Jul;27(4):795-800.

Levy, H.; Diallo, S.; Tennant, S.M.; Livio, S.; Sow, S.O.; Tapia, M.; Fields, P.I.; Mikoleit, M.; Tamboura, B.; Kotloff, K.L.; Lagos, R.; Nataro, J.P.; Galen, J.E. & Levine, M.M. (2008). PCR method to identify *Salmonella enterica* serovars Typhi, Paratyphi A, and

Paratyphi B among *Salmonella* Isolates from the blood of patients with clinical enteric fever. J Clin Microbiol. 2008 May;46(5):1861-6.

Ley, B.; Thriemer, K.; Ame, S.M.; Mtove, G.M.; von Seidlein, L.; Amos, B.; Hendriksen, I.C.; Mwambuli, A.; Shoo, A.; Kim, D.R.; Ochiai, L.R.; Favorov, M.; Clemens, J.D.; Wilfing, H.; Deen, J.L. & Ali, S.M. (2011). Assessment and comparative analysis of a rapid diagnostic test (Tubex®) for the diagnosis oftyphoid fever among hospitalized children in rural Tanzania. BMC Infect Dis. 2011 May 24;11:147.

Lim, P.L.; Tam, F.C.; Cheong, Y.M. & Jegathesan, M. (1998). One-step 2-minute test to detect typhoid-specific antibodies based on particle separation in tubes. J Clin Microbiol1998;36:2271-8.

Mahle, W.T. & Levine, M.M. (1993). *Salmonella* Typhi infection in children younger than five years of age. Pediatr Infect Dis J. 1993Aug;12(8):627-31.

Maskey, A.P.; Day, J.N.; Phung, Q.T.; Thwaites, G.E.; Campbell, J.I.; Zimmerman, M. Farrar, J.J. & Basnyat, B. (2006). *Salmonella enterica* serovar Paratyphi A and *S. enterica* serovar Typhi cause indistinguishable clinical syndromes in Kathmandu, Nepal. Clin Infect Dis. 2006; 42:1247-1253.

Massi, M. N.; Shirakawa, T.; Gotoh, A.; Bishnu, A.; Hatta, M. & Kawabata, M. (2003). Rapid diagnosis of typhoid fever by PCR assay using one pair of primers from flagellin gene of *Salmonella* Typhi. J Infect Chemother. 2003 Sep;9(3):233-7.

Massi, M.N.; Shirakawa, T.; Gotoh, A.; Bishnu, A.; Hatta, M. & Kawabata, M. (2005). Quantitative detection of *Salmonella enterica* serovar Typhi from blood of suspected typhoid fever patients by real-time PCR. Int J Med Microbiol. 2005 Jun;295(2):117-20.

Mathai, E.; John, T.J.; Rani, M.; Mathai, D.; Chacko, N.; Nath, V. & Cherian, A.M. (1995). Significance of *Salmonella* Typhi bacteriuria. J Clin Microbiol. 1995 Jul;33(7):1791-2.

Mazzulli, T.; Low, D.E. & Poutanen SM. (2005). Proteomics and severe acute respiratory syndrome (SARS): emerging technology meets emerging pathogen. Clin Chem. 2005 Jan;51(1):6-7.

Meltzer, E.; Sadik, C. & Schwartz, E. (2005). Enteric fever in Israeli travelers: a nationwide study. J Travel Med 2005; 12:275-281.

Mohanty, S.; Bakshi, S., Gupta, A.K.; Kapil, A.; Arya, L.S. & Das, B.K. (2003). Venousthrombosis associated with *Salmonella*: report of a case and review of literature. Indian J Med Sci. 2003; 57: 199-203.

Morin, N.J.; Gong, Z. & Li, X.F. (2004). Reverse transcription-multiplex PCR assay for simultaneous detection of Escherichia coli O157:H7, Vibrio cholerae O1, and *Salmonella* Typhi. Clin Chem. 2004 Nov;50(11):2037-44.

Moriñigo, M.A.; Muñoz, M.A.; Martinez-Manzanares, E.; Sánchez, J.M. & Borrego, J.J. (1993). Laboratory study of several enrichment broths for the detection of *Salmonella* spp. particularly in relation to water samples. J Appl Bacteriol. 1993 Mar;74(3):330-5.

Murray, P.R.; Spizzo, A.W. & Niles, A.C. (1991). Clinical comparison of the recoveries of bloodstream pathogens in Septi-Chek brain heart infusion broth with saponin, Septi-Chek tryptic soy broth, and the isolator lysis-centrifugation system. J Clin Microbiol. 1991 May;29(5):901-5.

Nagarajan, A.G.; Karnam, G.; Lahiri, A.; Allam, U.S. & Chakravortty, D. (2009). Reliable means of diagnosis and serovar determination of blood-borne *Salmonella* strains: quick PCR amplification of unique genomic loci by novel primer sets. J Clin Microbiol. 2009 Aug;47(8):2435-41.

Nair, S.; Lin, T.K.; Pang, T. & Altwegg, M. (2002). Characterization of *Salmonella* serovars by PCR-single-strand conformation polymorphism analysis. J Clin Microbiol. 2002 Jul;40(7):2346-51.

Nakhla, I.; El Mohammady, H.; Mansour, A.; Klena, J.D.; Hassan, K.; Sultan, Y.; Pastoor, R., Abdoel, T.H. & Smits, H. (2011). Validation of the Dri-Dot Latex agglutination and IgM lateral flow assays for the diagnosis of typhoid fever in an Egyptian population. Diagn Microbiol Infect Dis. 2011 Aug;70(4):435-41.

Nandagopal, B.; Sankar, S.; Lingesan, K.; Appu, K.C.; Padmini, B.; Sridharan, G. & Gopinath, A.K. (2010). Prevalence of *Salmonella* Typhi among patients with febrile illness in rural and peri-urban populations of Vellore district, as determined by nested PCR targeting the flagellin gene. Mol Diagn Ther. 2010 Apr 1;14(2):107-12.

Narayanappa, D.; Sripathi, R.; Jagdishkumar, K. & Rajani, H.S. (2010). Comparative study of dot enzyme immunoassay (Typhidot-M) and Widal test in the diagnosis of typhoid fever.Indian Pediatr. 2010 Apr;47(4):331-3.

Nath, G.; Mauryal, P., Gulati, A.K.; Singh, T.B.; Srivastava, R.; Kumar, K. & Tripathi, S.K. (2010). Comparison of Vi serology and nested PCR in diagnosis of chronic typhoid carriers in two different study populations in typhoid endemic area of India. Southeast Asian J Trop Med Public Health. 2010 May;41(3):636-40.

Ngan, G.J.; Ng, L.M.; Lin, R.T. & Teo, J.W. (2010). Development of a novel multiplex PCR for the detection and differentiation of *Salmonella enterica* serovars Typhi and Paratyphi A. Res Microbiol. 2010 May;161(4):243-8.

Nga, T.V.; Karkey, A.; Dongol, S.; Thuy, H.N.; Dunstan, S.; Holt, K.; Tu le, T.P.; Campbell, J.I.; Chau, T.T.; Chau, N.V.; Arjyal, A.; Koirala, S.; Basnyat, B.; Dolecek, C.; Farrar, J. & Baker, S. (2010). The sensitivity of real-time PCR amplification targeting invasive *Salmonella* serovars in biological specimens. BMC Infect Dis. 2010 May 21;10:125.

Nizami, S.Q.; Bhutta, Z.A.; Siddiqui, A.A. & Lubbad, L. (2006). Enhanced detection rate of typhoid fever in children in a periurban slum in Karachi, Pakistan using polymerase chain reaction technology. Scand J Clin Lab Invest. 2006;66(5):429-36.

Notomi, T.; Okayama, H.; Masubuchi, H.; Yonekawa, T.; Watanabe, K.; Amino, N. & Hase, T. (2000). Loop-mediated isothermal amplification of DNA. Nucleic Acids Res. 2008; 28: e63.

Nordmann, P.; Lartigue, M.F. & Poirel, L. (2008). Beta-lactam induction of ISEcp1B-mediated mobilization of the naturally occurring bla(CTX-M) beta-lactamase gene of Kluyvera ascorbata. FEMS Microbiol Lett. 2008 Nov;288(2):247-9.

Ochiai, R.L. Wang, X.; von Seidlein, L.; Yang, J.; Bhutta, Z.A.; Bhattacharya, S.K.; Agtini, M.; Deen, J.L.; Wain, J.; Kim, D.R.; Ali, M.; Acosta, C.J.; Jodar, L. & Clemens, J.D. (2005). *Salmonella* Paratyphi A rates, Asia. Emerg Infect Dis. 2005 Nov;11(11):1764-6.

Olopoenia, L.A. & King, A.L. (2000). Widal agglutination test - 100 years later: still plagued by controversy. Postgrad Med J. 2000 Feb;76(892):80-4.

Olsen, S.J.; Pruckler, J.; Bibb, W.; Nguyen, T.M.; Tran, M.T. & Nguyen, T.M.; Sivapalasingam, S.; Gupta, A.; Phan, T.P.; Nguyen, T.C.; Nguyen, V.C.; Phung, D.C. & Mintz, E.D. (2004). Evaluation of rapid diagnostic tests for typhoid fever. J Clin Microbiol. 2004 May;42(5):1885-9.

Omuse, G.; Kohli, R. & Revathi, G. (2010). Diagnostic utility of a single Widal test in the diagnosis of typhoid fever at Aga Khan University Hospital (AKUH), Nairobi, Kenya. Trop Doct. 2010 Jan;40(1):43-4.

Oracz, G.; Feleszko, W.; Golicka, D.; Maksymiuk, J.; Klonowska, A. & Szajewska, H. (2003) Rapid diagnosis of acute *Salmonella* gastrointestinal infection. Clin Infect Dis. 2003 Jan 1;36(1):112-5.

Ou, H.Y.; Ju, C.T.; Thong, K.L.; Ahmad, N.; Deng, Z.; Barer, M.R. & Rajakumar, K. (2007). Translational genomics to develop a *Salmonella enterica* serovar Paratyphi A multiplex polymerase chain reaction assay. J Mol Diagn. 2007 Nov;9(5):624-30.

Pandit, A.; Arjyal, A.; Paudyal, B.; Campbell, J.C.; Day, J.N.; Farrar, J.J. & Basnyat, B. (2008). A patient with paratyphoid A fever: an emerging problem in Asia and not always a benign disease. J Travel Med. 2008 Sep-Oct;15(5):364-5.

Palit, A.; Ghosh, S.; Dutta, S.; Sur, D.; Bhattacharya, M.K. & Bhattacharya, S.K. (2006). Increasing prevalence of *Salmonella enterica* serotype Paratyphi-A in patients with enteric fever in a periurban slum setting of Kolkata, India. Int J Environ Health Res. 2006 Dec;16(6):455-9.

Parry, C.M.; Hien, T.T.; Dougan, G.; White, N.J. & Farrar, J.J. (2002).Typhoid fever. N Engl J Med. 2002 Nov 28;347(22):1770-82.

Parry, C.M.; Wijedoru, L.; Arjyal, A. & Baker, S. (2011). The utility of diagnostic tests for enteric fever in endemic locations. Expert Rev Anti Infect Ther. 2011 Jun;9(6):711-25.

Patel, T.A.; Armstrong, M.; Morris-Jones, S.D.; Wright, S.G. & Doherty, T. (2010). Imported enteric fever: case series from the hospital for tropical diseases, London, United Kingdom. Am J Trop Med Hyg. 2010 Jun;82(6):1121-6.

Pokharel, B.M.; Koirala, J.; Dahal, R.K.; Mishra, S.K.; Khadga, P.K. & Tuladhar, N.R. (2006). Multidrug-resistant and extended-spectrum beta-lactamase (ESBL)-producing *Salmonella enterica* (serotypes Typhi and Paratyphi A) from blood isolates in Nepal: surveillance of resistance and a search for newer alternatives. Int J Infect Dis. 2006 Nov;10(6):434-8.

Prakash, P.; Mishra, O.P., Singh, A.K.; Gulati, A.K. & Nath, G. (2005). Evaluation of nested PCR in diagnosis of typhoid fever. J Clin Microbiol. 2005 Jan;43(1):431-2.

Rajagopal, A.; Ramasamy, R.; Mahendran, G. & Thomas, M. (2002). Hepatic abscess complicating paratyphoid infection. Trop Gastroenterol. 2002; 23: 181–82.

Reynolds, D.W.; Carpenter, R.L. & Simon, W.H. (1970). Diagnostic specificity of Widal's reaction for typhoid fever. JAMA. 1970 Dec 21;214(12):2192-3.

Rello, J.; Lisboa, T.; Lujan, M.; Gallego, M.; Kee, C.; Kay, I.; Lopez, D.; Waterer, G.W. & DNA-Neumococo Study Group. (2009).Severity of pneumococcal pneumonia associated with genomic bacterial load. Chest. 2009 Sep;136(3):832-40.

Rubin, F.A.; Kopecko, D.J.; Sack, R.B.; Sudarmono, P.; Yi, A.; Maurta, D.; Meza, R.; Moechtar, M.A.; Edman, D.C. & Hoffman, S.L. (1988). Evaluation of a DNA probe for identifying *Salmonella* Typhi in Peruvian and Indonesian bacterial isolates. J Infect Dis. 1988 May;157(5):1051-3.

Rubin, F.A.; McWhirter, P.D.; Punjabi, N.H.; Lane, E.; Sudarmono, P.; Pulungsih, S.P.; Lesmana, M.; Kumala, S.; Kopecko, D.J. & Hoffman, S.L. (1989). Use of a DNA probe to detect *Salmonella* Typhi in the blood of patients with typhoid fever. J Clin Microbiol. 1989 May;27(5):1112-4.

Rubin, F.A.; McWhirter, P.D.; Burr, D.; Punjabi, N.H.; Lane, E.; Kumala, S.; Sudarmono, P.; Pulungsih, S.P.; Lesmana, M. & Tjaniadi, P. (1990). Rapid diagnosis of typhoid fever through identification of *Salmonella* Typhi within 18 hours of specimen acquisition by culture of the mononuclear cell-platelet fraction of blood. J Clin Microbiol. 1990 Apr;28(4):825-7.

Sánchez-Jiménez, M.M. & Cardona-Castro, N. (2004). Validation of a PCR for diagnosis of typhoid fever and salmonellosis by amplification of the *hil*A gene in clinical samples from Colombian patients. J Med Microbiol. 2004 Sep;53(Pt 9):875-8.

Sinha, A.; Sazawal, S.; Kumar, R.; Sood, S.; Reddaiah, V.P.; Singh, B.; Rao, M.; Naficy, A.; Clemens, J.D. & Bhan, M.K. 1999. Typhoid fever in children aged less than 5 years.Lancet. 1999 Aug 28;354(9180):734-7.

Smith, S.I.; Bamidele, M.; Fowora, M.; Goodluck, H.T.; Omonigbehin, E.A.; Akinsinde, K.A.; Fesobi, T.; Pastoor, R.; Abdoel, T.H. & Smits, H.L. (2011). Application of a point-of-care test for the serodiagnosis of typhoid fever in Nigeria and the need for improved diagnostics. J Infect Dev Ctries. 2011 Jul 27;5(7):520-6.

Snyder, G.E.; Shaps, H.J. & Nelson, M. (2004). Multiple organ dysfunctionsyndrome associated with *Salmonella* Typhi infection. Am J Emerg Med. 2004; 22: 138–39.

Snyder, M.J.; Hornick, R.bB.; Mccrumb, F.R. Jr.; Morse, L.J. & Woodward, T.E. (1963). asymptomatic typhoidal bacteremia in volunteers. antimicrob agents chemother (bethesda). 1963;161:604-7.

Song, J.H.; Cho, H.; Park, M.Y.; Na, D.S.; Moon, H.B. & Pai, C.H. (1993). Detection of Salmonella Typhi in the blood of patients withtyphoid fever by polymerase chain reaction. J Clin Microbiol. 1993 Jun;31(6):1439-43.

Soper, G.A. (1939). The Curious Career of Typhoid Mary. Bull N Y Acad Med. 1939 Oct;15(10):698-712.

Stuart, B.M. & Pullen, R.L. (1946). Typhoid fever: clinical analysis of three hundred and sixty cases. Arch Intern Med. 1946;78:629–661.

Sztein, M.B. (2007). Cell-mediated immunity and antibody responses elicited by attenuated *Salmonella enterica* serovar Typhi strains used as live oral vaccines in humans. Clin Infect Dis. 2007 Jul 15;45 Suppl 1:S15-9.

Tam, F.C. & Lim, P.L. (2003). The TUBEX typhoid test based on particle-inhibition immunoassay detects IgM but not IgG anti-O9 antibodies. J Immunol Methods. 2003 Nov;282(1-2):83-91.

Tam, F.C.; Wang, M.; Dong, B.; Leung, D.T.; Ma, C.H. & Lim, P.L. (2008). New rapid test for paratyphoid a fever: usefulness, cross-detection, and solution. Diagn Microbiol Infect Dis. 2008 Oct;62(2):142-50.

Teh, C.S.; Chua , K.H.; Puthucheary, S.D. & Thong, K.L. (2008). Further evaluation of a multiplex PCR for differentiation of *Salmonella* Paratyphi A from other *Salmonellae*. Jpn J Infect Dis. 2008 Jul;61(4):313-4.

Thisyakorn, U.; Mansuwan, P. & Taylor, D.N. (1987). Typhoid and paratyphoid fever in 192 hospitalized children in Thailand. Am J Dis Child. 1987 Aug;141(8):862-5.

Vallenas, C.; Hernandez, H.; Kay, B.; Black, R. & Gotuzzo, E. (1985). Efficacy of bone marrow, blood, stool and duodenal contents cultures for bacteriologic confirmation of typhoid fever in children. Pediatr Infect Dis. 1985 Sep-Oct;4(5):496-8.

Vincent, M.; Xu, Y. & Kong, H. (2004). Helicase-dependent isothermal DNA amplification. EMBO reports, 2004; 5: 795–800.

Volk, E.E.; Miller, M.L.; Kirkley, B.A. & Washington, J.A. (1998). The diagnostic usefulness of bone marrow cultures in patients with fever of unknown origin. Am J Clin Pathol. 1998 Aug;110(2):150-3.

Vollaard, A.M.; Ali, S.; Widjaja, S.; Asten, H.A.; Visser, L.G.; Surjadi.; C. & van Dissel, J.T. (2005). Identification of typhoid fever and paratyphoid fever cases at presentation in outpatient clinics in Jakarta, Indonesia. Trans R Soc Trop Med Hyg. 2005; 99:440–450.

Wain, J. & Hosoglu, S. (2008). The laboratory diagnosis of enteric fever. *J Infect Dev Ctries.* 2008 Dec 1;2(6):421-5.

Wain, J.; Diep, T.S.; Ho, V.A.; Walsh, A.M.; Nguyen, T.T.; Parry, C.M. & White, N.J. (1989). Quantitation of bacteria in blood of typhoid fever patients and relationship between counts and clinical features, transmissibility, and antibiotic resistance. J Clin Microbiol. 1998 Jun;36(6):1683-7.

Wain, J.; House, D.; Zafar, A.; Baker, S.; Nair, S.; Kidgell, C.; Bhutta, Z.; Dougan, G. & Hasan, R. (2005). Vi antigen expression in *Salmonella enterica* serovar Typhi clinical isolates from Pakistan. J Clin Microbiol. 2005 Mar;43(3):1158-65.

Wain, J.; Pham, B.V.; Ha, V.; Nguyen, N.M.; To, S.D.; Walsh, A.L.; Parry, C.M.; Hasserjian, R.P.; HoHo, V.A.; Tran, T.H.; Farrar, J.; White, N.J. & Day, N.P. (2001). Quantitation of bacteria in bone marrow from patients with typhoid fever: relationship between counts and clinical features. J Clin Microbiol. 2001 Apr;39(4):1571-6.

Wain, J.; To, S.D.; Phan, V.B.B.; Walsh, A.L.; Ha, V.; Nguyen, M.D.; Vo, A.H.; Tran, T.H.; Farrar, J.; White, N.J.; Parry, M. & Day, N.P.J. (2008). Specimens and culture media for the laboratory diagnosis of typhoid fever. *J Infect Dev Ctries.* 2008 Dec 1;2(6):469-74.

Walker, G.T.; Fraiser, M.S.; Schram, J.L.; Little, M.C.; Nadeau, J.G. & Malinowski, D.P. (1992). Strand-displacement amplification-an isothermal, in vitro DNA amplification technique. Nucleic Acids Res. 1992; 20: 1691–1696.

Watson, K. C. (1955). Isolation of *Salmonella* Typhi from the blood stream. J Lab Clin Med. 1955 Jul;46(1):128-34.

Whitaker, J.A.; Franco-Paredes, C.; del Rio, C. & Edupuganti, S. (2009). Rethinking typhoid fever vaccines: implications for travelers and people living in highly endemic areas. J Travel Med. 2009 Jan-Feb;16(1):46-52.

Widal, F.M. (1896). Serodiagnostic de la fiévre typhoide a-propos d'uve modification par MMC Nicolle et al. Halipie. Bull Soc Med Hop Paris, 1896; 13:561–566.

Woods, C.W.; Murdoch, D.R.; Zimmerman, M.D.; Glover, W.A.; Basnyat, B.; Wolf, L.; Belbase, R.H. & Reller, L.B. (2006). Emergence of *Salmonella enterica* serotype Paratyphi A as a major cause of enteric fever in Kathmandu, Nepal.Trans R Soc Trop Med Hyg. 2006 Nov;100(11):1063-7.

World Health Organization. Background document: The diagnosis, treatment and prevention of typhoid fever. WHO/V&B/03.07.Geneva: World Health Organization, 2003.

Yang, H.H.; Gong, J.; Zhang, J.; Wang, M.L,.; Yang, J.; Wu, G.Z,.; Quan, W.L.; Gong, H.M. & Szu, S.C. (2010).An outbreak of *Salmonella* Paratyphi A in a boarding school: a community-acquired enteric fever and carriage investigation. Epidemiol Infect. 2010 Dec;138(12):1765-74.

Zhou, L. & Pollard, A.J. (2010). A fast and highly sensitive blood culture PCR method for clinical detection of *Salmonella enterica* serovar Typhi. Ann Clin Microbiol Antimicrob. 2010 Apr 19;9:14.

Zhu, Q.; Lim, C.K. & Chan, Y.N. (1996). Detection of *Salmonella* Typhi by polymerase chain reaction. J Appl Bacteriol. 1996 Mar;80(3):244-51.

4

Molecular Technologies for *Salmonella* Detection

Robert S. Tebbs, Lily Y. Wong, Pius Brzoska and Olga V. Petrauskene

Life Technologies, Foster City, CA
USA

1. Introduction

Salmonella has been associated with some of the most devastating foodborne outbreaks in recent history. *Salmonella* outbreaks have been linked to a variety of foods including produce [Alfalfa Sprouts- 2009, 2010, 2011; pistachios-2009; cantaloupes-2008, 2011 etc.], processed foods [peanuts – 2009], and prepared foods [turkey burgers- 2011, Banquet Pot Pies – 2007]. The contamination of commercial shell eggs with *Salmonella* Enteriditis in 2010 led to the recall of over a half a billion eggs, and the contamination of peanut-containing products with *Salmonella* Typhimurium in 2008-2009 led to one of the largest recalls in U.S. history with over 3,900 products being recalled. The Peanut Corporation of America, responsible for the *Salmonella* outbreak in peanuts, was forced into bankruptcy. Multiple lawsuits were filed against Wright County Egg and Hillandale Farms responsible for the *Salmonella* outbreak in eggs. Despite their own internal testing which showed *Salmonella* contamination, these facilities still shipped product. The 2008 outbreak of *Salmonella* in jalapeno peppers resulted in 1442 persons infected with *Salmonella* Saintpaul across 43 states, the District of Columbia, and Canada. Unfortunately, the tomato industry was implicated early in the investigation, which resulted in economic losses to the tomato industry in hundreds of millions of dollars. Because *Salmonella* is widespread in the environment (in such places as chicken houses), vegetable plants and animals (as well as meat samples, eggs etc.), rapid, reliable, and validated pathogen detection methods are needed for use in production facilities, public health labs, as well as in the regulatory and monitoring agencies. To provide comprehensive rapid food testing solutions, all components of a pathogen detection system should be addressed: sample preparation, detection and data analysis.

Fluorescent quantitative real-time PCR is the most sensitive method for detection, monitoring and measurement of pathogen levels. The method also can be used for strain identification based on single-nucleotide polymorphism detection. A key element in designing PCR assays is an algorithm to select primers and probes because they define accuracy – specificity and inclusivity of the PCR tests. The ability to design highly specific assays becomes easier as the number of bacterial genomes added to the public domain increases.

There are a number of sample preparation methods available that are fast and easy for PCR-based pathogen detection using both low throughput (manual) and high throughput (automated) methods.

A multiplex (multi-color) real-time PCR analysis, if designed correctly, provides simultaneous and specific detection of a number of pathogens in the same reaction and can save time and money. In addition to multiplexing, other technologies improve operator experience such as lyophilized configuration and fast cycling format.

Effective data analysis software can significantly improve test workflow as well as accuracy of the presence or absence calls. Software packages can simplify analysis by displaying results graphically to make the system fail-proof even for novice users.

To further characterize detected *Salmonella* species, isolates can be sequenced using modern whole genome sequencing platforms.

2. Genetic methods complement biochemical and phenotypic analyses

Salmonella serotypes are classified by the Kauffmann-White-Schema which is maintained by the WHO and the Collaborating Center for Reference and Research at Institute Pasteur. There are only two species *Salmonella enterica*, which is associated with human infections, and *Salmonella bongori*, which is mainly found in lizards. This schema was based mainly on DNA analysis and bonified by the judiciary of bacterial nomenclature (Center for Disease Control, 2004). Under the schema, there are six subspecies. The serotypes are I: *Salmonella enterica* subsp enterica, II: *Salmonella enterica* subsp. Salamae, III: *Salmonella enterica* subsp. Arizonae, IIIb *Salmonella enterica* subsp. Diarizonae, IV: *Salmonella enterica* subsp. Houtenae and VI *Salmonella enterica* susp. Indica. *Salmonella* serotyping is traditionally based on immunoreactive antibodies against the O and H antigens. Different classification schemes based on phylogenetic analyses of 16S and housekeeping genes have been proposed (Boyd et al., 1996; Tindall, 2005).

Classification of bacteria is traditionally based on immunogenic and metabolic behavior. Analysis of bacterial genomes however led to reclassification and debates on the taxonomical classification. For example, Pupo et al. (2000) studied the phylogenetic relations of several housekeeping genes and the O-antigen of species of the genus *Shigella* and concluded that several species of *Shigella* are clusters of *Escherichia coli*.

The nomenclature change in *Salmonella* was subsequently supported by genomic evidence as well. McQuiston et al. (2008) showed that a set of four housekeeping genes supports the *Salmonella* classification, and microarray analysis of the gene homologues in *Salmonella* result in a similar grouping (Porwollik et al., 2002). More recently, whole genome sequencing of *Enterobacter sakazakii* isolates revealed that this group is phylogenetically different from other *Enterobacter* species and was renamed to *Cronobacter sakazakii* (Iversen et al., 2008). The breakout of *Cronobacter* was supported by biochemical and microarray analyses (Healy et al., 2009).

Since the completion of the first bacterial genome of Hemophilus influenza in 1995 (Fleischmann et al., 1995), more than 1000 bacterial genomes have been completely sequenced. Currently, 21 serovars of *Salmonella enterica* subsp. enterica have been sequenced as well as *Salmonella bongori*. Many shotgun sequencing projects are still in progress, and the number of genomes will continue to increase. New metrics for taxonomical evaluation based on complete genomes have been proposed (Kunin et al., 2005). Complete genomic information does not change the phylogeny based on 16S and

MLST substantially (Coenye & Vandamme, 2003), but it allows the study of specific genes present and absent across phylogenetic groups.

3. Sample preparation for real-time PCR detection of *Salmonella*

The successful detection of pathogenic organisms by genetic methods requires microbial lysis to release nucleic acids and efficient removal of inhibitors. Sample preparation can also serve to concentrate nucleic acids for improved sensitivity. Food and environmental samples create unique challenges for sample preparation due to the heterogeneous nature of the different matrices. The method used must account for the type and amount of organism to be lysed, the sample matrix, and the user's needs and limitations (cost, ease-of-use, time-to-results, sample throughput and capacity, and multi-functionality).

Samples that contains inhibitory compounds can lead to partial or complete inhibition of PCR. Food and culture media both contain components that can inhibit PCR (Rosen et al., 1992; Andersen & Omiecinski, 1992; Atmar et al., 1993; Demeke & Adams, 1992; Lofstrom et al., 2004) (for a review, see Wilson, 1997). PCR inhibitors originating from the food samples include humic acid from soil (Tsai & Olson, 1992a; Tsai & Olson, 1992b), proteins and aminoglycans from animal samples such as hemoglobin, lactoferrin and heparin (Al-Soud & Radstrom, 2001), polysaccharides from plant material (Demeke & Adams, 1992; Monteiro et al., 1997), melanin from hair and skin (Eckhart et al., 2000), etc. Media including modified Rappaport broth and phosphate buffered saline can inhibit PCR (Rossen et al., 1992). PCR can also be inhibited by contaminants from the nucleic acid extraction phase including ionic detergents (Weyant et al., 1990), phenol, ethanol, proteinase K, guanidinium, and salts (Al-Soud & Radstrom, 2001).

The control of PCR inhibition can be addressed on several fronts. Inhibitory effects can be minimized by optimizing the PCR mix. Bovine serum albumin (BSA) was shown to reduce PCR inhibition by humic acid (McGregor et al., 1996) and hemoglobin (Al-Soud & Radstrom, 2001), possibly through direct interaction with the inhibitory components such that they cannot interfere with PCR amplification. The single-stranded DNA binding protein from bacteriophage T4 (gp32) also reduced PCR inhibition caused by hemoglobin (Al-Soud & Radstrom, 2001). The addition of Tween® 20 or DMSO reversed PCR inhibition from low concentrations of the polysaccharides dextran sulfate and gum ghatti (Demeke & Adams, 1992). Inhibitors that affect polymerase activity can be partially mitigated by increasing the polymerase concentration. PCR kit manufacturers often develop proprietary formulations to optimize PCR through design of experiment (DOE) studies. For example, Environmental Master Mix version 2 (EMMv2) was developed by Applied Biosystems specifically for complex samples containing potential inhibitory components (Figure 1). PCR inhibition can be monitored using an internal positive control (IPC) (Tebbs et al., 2010). Samples that show no amplification of target and IPC either contain inhibitors or the PCR reaction was improperly prepared. Technical errors are greatly minimized with new lyophilized formulations that only require the addition of sample. Amplification of the internal control gives confidence that a negative result is not due to inhibition. A simple mitigation to inhibited samples is dilution (Tsai & Olson, 1992a; Tsai & Olson, 1992b). For samples in which the target DNA is not the limiting factor, inhibitors can be diluted below their effective threshold to allow for PCR amplification. In addition, efficient DNA extraction following bacterial enrichment removes PCR inhibitors and improves accurate detection.

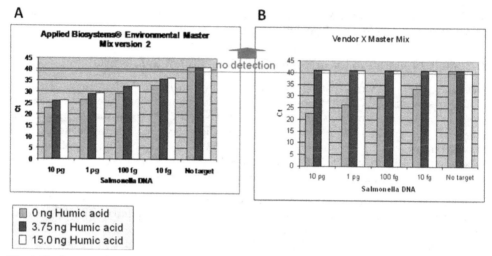

Fig. 1. Real-time PCR Master Mixes can mitigate PCR inhibition. (A) Environmental Master Mix version 2 shows detection of *Salmonella* DNA in the presence of 3.75 ng and 15.0 ng of humic acid. (B) Master mix from another source shows complete inhibition in the presence of the same amount of humic acid. Real-time PCR was performed on the 7500Fast instrument.

Food borne pathogens are usually present in small quantities in food and therefore require enrichment to detect their presence. The United States requires testing for selected pathogens in the nation's food supply. This is true for *Salmonella* species in foods such as ready-to-eat products and whole shell eggs. A standard practice for screening food for regulated pathogens is to mix 25 grams of food with 225 mL of broth (1:9 food to broth ratio). Reference methods are based on traditional culture procedures and typically use a 2-step enrichment procedure, first in non-selective broth (pre-enrichment) and second in selective broth, prior to biochemical and serological characterization. These protocols are designed to detect down to a single viable organism. The pre-enrichment step allows recovery of injured or otherwise weakened *Salmonella*, whereas selective enrichment favors growth of *Salmonella* over background flora that competes with *Salmonella* for available nutrients. The U.S. FDA *Bacteriological Analytical Manual* (BAM) for *Salmonella* pre-enriches in different broths depending on the food matrix (typically lactose broth or tryptic soy broth), followed by selective enrichment in Rappaport-Vassiliadis (RV) medium and tetrathionate (TT) broth (Andrews & Hammack, 2011). Pre-enrichment is for 24 ± 2 h at 35°C, and selective enrichment is for 24 ± 2 h at 42°C for RV and 35°C (food with low microbial load) or 43°C (food with high microbial load) for TT. The U.S. FDA protocol for sampling and detecting *Salmonella* in poultry houses pre-enriches environmental samples in buffered peptone water (24 ± 2 h at 35°C) and then selectively enriches in RV (24 ± 2 h at 42°C) and TT (24 ± 2 h at 43°C) (Food and Drug Administration, 2008). The International Organization for Standardization (ISO) reference method for the detection of *Salmonella* in food (ISO 6579:2002(E)) recommends pre-enrichment in buffered peptone water (other broths are necessary for some food types) for 18 ± 2 h at 37°C followed by selective enrichment in Rappaport-Vassiliadis medium with soya (RVS broth) for 24 ± 3 h at 41.5°C and Muller-Kauffman tetrathionate/novobiocin broth (MKTTn broth) for 24 ± 3 h at 37°C

(ISO 6579:2002 [E]). The enrichment time alone for these traditional culture methods totals 2 days. Following enrichment, the samples are plated to selective agar plates for 24-48 h growth and then transferred to slants for another 24 h growth. Presumptive detection of *Salmonella* is determined in 4 to 5 days (BS plates are left for 48 ± 2 h for BAM method).

Food producers desire faster time-to-results since it provides great cost benefits. The Food and Safety Inspection Service (FSIS) requests all meat and poultry products be held by producers until test result indicate no pathogen is present. The FSIS recently proposed new regulations requiring test results be received before meat and poultry can be shipped, the so called "Test-and Hold" policy (Department of Agriculture, 2011). To the producer, storing product is not only costly but also shortens the product shelf life.

Rapid methods can shorten time-to-results to less than 24 h for *Salmonella*. To be adopted by the food industry, new methods must undergo AOAC validation to demonstrate equivalency to reference culture methods. The AOAC developed the Performance Tested Methods[SM] (PTM) program for the purpose of certifying commercial test kits (AOAC Research Institute website). Even with AOAC validation demonstrating equivalency, the FDA BAM considers positive results from rapid methods to be presumptive that must either be confirmed by culture or accepted as true positive (negative samples are accepted as true negative). Rapid methods are fast because detection is immediate, but also because enrichment requirements are usually shorter. Most PCR methods only require a single-step enrichment to demonstrate equivalency to reference standards. Well designed PCR assays can detect a single genomic copy of *Salmonella* and thus the limit of detection for PCR is largely determined by sample preparation.

Multiple methods have been used to lyse pathogens including physical, chemical and enzymatic, or combinations of the three (Table 1). Common physical methods include temperature (freeze/thaw or heat), bead-beating, and sonication. Freeze/thaw lysis is a traditional method in which the cellular suspension is transferred between freezing and warm conditions, for example between a dry ice-ethanol bath and a 37°C water bath. During the freeze cycle ice crystals cause cells to expand and rupture. Multiple freeze/thaw cycles are required for efficient lysis which makes the procedure rather lengthy and is usually only associated with "home brew" methods. Microbial cells can also be lysed by heating at 95°C to 100°C. The lysis efficiency of heat is dependent on the microorganism, but is generally poor. Heat is often combined with chemical and/or enzymatic treatment to increase the lysis efficiency which is discussed later. If enzymatic treatment is used, then the heat step serves two functions, it can break open cells and inactivate enzymatic activity. Protease is common for bacterial lysis and if used must be inactivated before adding sample to a PCR mix since proteases will destroy DNA polymerase. Sonication uses high frequency sound waves to create localized regions of low pressure resulting in micro bubbles that rapidly form and implode, ultimately breaking open cells. Bead-beating is another physical method used for breaking open cells. Typically an equal volume of silica or zirconium beads (approximately 0.1 mm diameter) are combined with a sample and mixed on a laboratory vortex. Lysis is complete in 3-5 minutes. Bead-beating has a tendency to generate foam which can be controlled by using anti-foam agents. Bead-beating and sonication can result in greatly fragmented, low molecular weight nucleic acids. However, fragmentation is of minor concern for real-time PCR since assays are designed to amplify small fragments of typically less than 100 base pairs.

Lysis Method	Category	Product	Notes	Test Kit Name	Manufacturer
Bead-Beating/ Chemical/Filter	Manual	DNA		IT 1-2-3™ Sample Purification Kits	Idaho Technology
Chemical	Automated	Lysate	DNA Hybridization	GeneQuence® *Salmonella*	Neogen
Chemical/ Enzyme/ Heat	Simple high throughput	Lysate	2-step enrich (except meat and poultry)	BAX® System PCR Assay *Salmonella*	DuPont Qualicon
Chemical/Heat	Simple	Lysate		PrepSEQ® Rapid Spin Sample Preparation Kit	Life Technologies
Chemical/Heat	Simple	Lysate	2-step enrich	Foodproof® ShortPrep I Kit	BIOTECON Diagnostics
Chemical/Heat	Simple	Lysate		iQ-Check Salmonlla II Easy Extraction I	Bio-Rad
Chemical/Heat	Simple high throughput	Lysate	96-well Deepwell centrifugation	iQ-Check® *Salmonella* II Deepwell protocol	Bio-Rad
Chemical/Heat/ Magnetic beads	Automated	DNA		Foodproof® Magnetic Preparation Kit I	BIOTECON Diagnostics
Chemical/Heat/ Magnetic beads	Semi-automated	DNA/ RNA		PrepSEQ® Nucleic Acid Extraction Kit	Life Technologies
Chemical/Heat/ Filter	Manual	DNA		Foodproof® Sample Preparation Kit	BIOTECON Diagnostics
Chemical/Heat/ Filter	Manual	DNA		Biotest MMB Prep *Salmonella*	Biotest AG
Chemical/Heat/ Filter	Manual	DNA		SureFood® Prep *Salmonella*	Congen Biotechnologie
Sonication/Heat	Automated	DNA		GeneDisc® *Salmonella*	Pall Corp.

Table 1. Sample preparation kits for detecting *Salmonella* by genetic methods. The kits included were chosen from the AOAC Research Institute online website of Performance Tested Methods[SM] Validated for detection of *Salmonella* using genetic methods (http://www.aoac.org/testkits/testedmethods.html). Details for sample preparation were obtained from readily available information from company websites and might not be part of the AOAC approved workflow. AES and BioControl have AOAC validation for real-time PCR detection of *Salmonella,* but were omitted from the current list since details of their sample preparation methods were not available on their website.

Many chemicals have been used in cell lysis. The most common chemicals are detergents. Detergents disrupt the lipid bilayer surrounding cells. There are many types of detergents of varying strengths. Ionic detergents such as sodium dodecyl sulfate (SDS) are stronger than nonionic or zwitterionic detergents and are often employed for microbial lysis to extract DNA. In addition to disrupting lipids, SDS has the advantage of denaturing proteins, including DNAse and RNAse, and thus protecting nucleic acid during extraction. Because detergents cannot lyse cell walls, bacteria are often pre-treated with enzymes (proteases or lysozyme) before addition of detergent. Chaotropic salts are also commonly used for bacterial lysis. Guanidinium thocyanate and guanidinium chloride lyse cell membranes by denaturing proteins. Chemicals used in bacterial lysis are by nature hazardous and must be disposed of as hazardous waste.

The use of magnetic particles for sample preparation has increased in recent years. Imunomagnetic separation (IMS) uses metal beads coated with antibodies specific to the target microbe of interest. A magnet can then be used to attract the bead containing the target microbe. The beads can be washed, and the presence of the target organism can be determined by plating onto selective agar, PCR, or other detection methods. Another solution is to use solid phase capture in which antibodies are linked to a solid support. The target microbe can be captured as sample passes across the support. For example, antibodies linked to a pipette tip can capture the target organism by collecting sample with a pipetting device. In theory, the target analyte binds while unwanted material passes through. The analyte can be further purified through a series of wash steps. In practice, antibodies typically show background capture of non-target organisms (Fratamico et al., 2011), and therefore the use of IMS is often combined with a detection method such as PCR or biochemical characterization. Furthermore, the sample matrix (i.e. high fat content) can affect binding of an antibody to its antigenic substrate (Bosilevac et al., 2010; Fitzmaurice, 2006). Both false positive and false negative results have been reported for antibody-based methods. Both immunomagnetic beads and solid support systems have been used in combination with antibodies or phage-binder proteins to capture microbes. Phage-binder proteins are the proteins responsible for phage binding to host bacteria and are very host specific. Both technologies (beads and solid support) have also been used to capture nucleic acids from microbial lysates. Metal beads coated with silica and glass fiber filters are commonly used to capture nucleic acids. DNA binds under conditions of salt and high alcohol, and is eluted with aqueous solutions. Heating the elution buffer can improve recovery by increasing the ability of the nucleic acid to dissolve in solution. The advantage of these systems is increased target concentration and improved target purity without the need for centrifugation and aspiration—methods that are difficult to automate. The use of magnetic beads to capture nucleic acids eliminates the concern associated with antibody specificity since total DNA and RNA can be captured indiscriminately. The same is true for purification columns that bind nucleic acids to silica membranes. The specificity of the assay is determined by the detection method. The binding capacity for nucleic acids can be much higher for magnetic particles compared to that of standard spin columns. The disadvantage with the capture of total nucleic acids is the potential for the capture of large amounts of non-targeted DNA relative to target DNA.

There are ways to simplify sample preparation. How to simplify depends on the needs of the end-user. If obtaining results within an 8-hour work shift is critical, then simplification becomes more challenging. Quick time-to-results requires concentrating the microbe from large sample volumes in combination with efficient microbial lysis and DNA/RNA recovery. To meet this demand sample preparation becomes more complex. Automation can simplify the process, but adds additional costs for equipment. A common practice is to use magnetic particles to capture microbes or microbial nucleic acid since magnetic particles can be easily added to and extracted from an aqueous mixture. Furthermore, procedures using magnetic particles are simple to automate. Instruments can be designed to dispense and aspirate liquids, but instruments that perform these functions are typically more complex.

Increased enrichment times have the advantage of increasing the concentration of the microbe which can greatly simplify sample preparation (Figure 2). Enrichment also dilutes dead cells that could be present in the sample matrix. A 2-step enrichment method (e.g. pre-enrichment followed by selective enrichment) will also dilute the sample matrix which can reduce the impact of inhibitory substances associated with food and environmental samples, thus creating a more consistent sample for sample preparation. Double-enrichment has not found favor in molecular methods that are expected to be rapid. Indeed, PCR is very sensitive and double-enrichment is excessive for most applications. Most food and environmental samples can be enriched overnight (16-18 h) in one broth to allow *Salmonella*, if present, to grow to concentrations that are above the detection limit of PCR even when simple sample preparation methods are used.

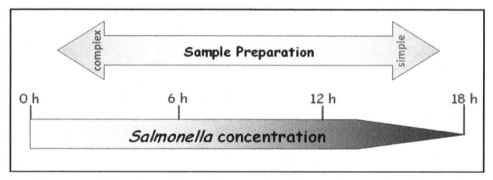

Fig. 2. Increasing *Salmonella* enrichment time can simplify sample preparation. Increased enrichment time increases *Salmonella* concentration requiring less volume and lower efficient sample preparation.

The simplest sample preparation methods dilute enriched samples into a solution that is compatible with the detection method. Indeed, many Gram-negative bacteria are lysed by boiling in water for 10 minutes. Because PCR amplification begins with a denaturation step (typically 95°C for 10 minutes) it is theoretically possible to add diluted sample directly to PCR. However, this doesn't seem to work well for many samples. Some level of lysis prior to adding to the PCR mix improves detection. For example, the addition of a simple 10 minute boiling step prior to setting up a PCR greatly improves detection. It is likely that boiling denatures enzymes and degrades substrates that disrupt the PCR reaction. A simple sample preparation method is shown in Figure 3.

A

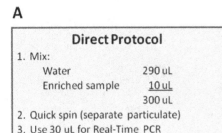

B

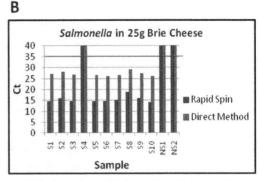

Fig. 3. Simple sample preparation for real-time PCR detection of *Salmonella*. (A) According to the direct protocol, sample is diluted in water and briefly centrifuged for 10 to 15 seconds in a table-top centrifuge. (B) Twenty-five grams of Brie cheese was spiked with 1-3 CFU *Salmonella enterica* serotype Typhimurium (strain Q210) and enriched at 37°C for 16 hours in buffered peptone water. The Direct Method showed a >10 Ct difference compared to the PrepSEQ® Rapid Spin method, but there was 100% correlation for detection of *Salmonella* between the two methods. Sample 4 (S4) which also received the spike gave negative results by both sample preparation methods, demonstrating all samples were spiked with low concentrations of *Salmonella* (i.e. fractional positive spike). Ten Cts correspond to a 1000-fold dilution in available DNA. Samples were analyzed on the 7500 Fast instrument using the MicroSEQ® *Salmonella* spp. Detection Kit.

4. Genetic-based methods for detection of *Salmonella* in foods

Genetic methods involve specific detection of RNA or DNA sequences to determine presence of the pathogen. There are a number of available kits in the market that apply genetic methods for the detection of *Salmonella* (Table 2). The most common genetic detection methods are PCR-based technologies. In the simplest form, conventional PCR involves amplification of a target DNA sequence using primers. The reaction is cycled between denaturing and annealing temperatures and may include a specific temperature for extension. The reaction generates an amplicon which can be detected on an agarose gel when stained with an intercalating dye such as ethidium bromide. The amplicon must be of the expected size for the sample to be called positive.

A DNA-binding dye, such as SYBR® Green, can be added to PCR and monitored by a real-time PCR instrument. SYBR® Green preferentially binds double-stranded DNA resulting in a DNA-Dye complex that shows a unique absorbance and emission spectrum. SYBR® Green dye will detect PCR products as they are amplified. Highly specific primer designs are required to avoid false positives because SYBR® Green dye will bind to all double-stranded DNA including any mis-primed products. Melt curve analysis of the PCR products of SYBR® Green dye reactions can be added to the end of a real-time PCR run to collect melting temperature (Tm) data of the PCR products amplified. This additional layer of data provides another check that the product amplified is of the expected Tm which is indicative of the fragment length. There are numerous examples of SYBR® Green PCR assays used to detect *Salmonella* in food or environmental samples (Nam et al., 2005; Techathuvanan et al., 2011).

Method	Test Kit Name	Manufacturer
PCR/melt curve analysis	BAX® System for *Salmonella*	DuPont Qualicon
Real-time PCR	Assurance GDS® for *Salmonella*	BioControl Systems
	Foodproof® *Salmonella* Detection Kit	BIOTECON Diagnostics
	GeneDisc® *Salmonella*	Pall Corp.
	iQ-Check® *Salmonella* II Kit	Bio-Rad
	MicroSEQ® *Salmonella* spp. Detection Kit	Life Technologies
	Salmonella species LT Test Kit	Idaho Technology
	SureFood® *Salmonella* PLUS V	Congen
DNA hybridization	GeneQuence® *Salmonella*	Neogen
NASBA	Nuclisens EasyQ® Basic Kit*	bioMerieux
LAMP	LoopAmp® DNA or RNA Amplification Kit*	Eiken Chemical Company

* General reagent kit which is not *Salmonella* specific.

Table 2. Commercial kits utilizing genetic methods to detect *Salmonella*.

Using fluorogenic probes with real-time PCR has rapidly become the standard for genetic detection because of its high specificity and sensitivity to detect low copy numbers. Many real-time instrument platforms have the ability to complete a run under one hour because of fast temperature ramping and improvements in master mix chemistries. TaqMan® assay uses target-specific primers and a probe that is labeled with a fluorescent reporter dye on the 5′ end and a quencher dye on the 3′ end (Tebbs et al., 2009; Balachandran et al., 2011). The probe anneals to the target DNA sequence between the two primer sites. As the primer extends during each cycle of the PCR, the 5′ nuclease activity of Taq polymerase displaces the probe from the DNA strand, separating the reporter dye and quencher dye in the process, and fluorescent signal is emitted. An example of a real-time assay which detects *Salmonella* Typhimurium is shown in Figure 4. Details of TaqMan data analysis will be further discussed.

An alternative fluorescent probe is the molecular beacon. The stem-and-loop structure of a molecular beacon probe consists of a target-specific sequence (which forms the loop) and non-target sequences that are complementary at the 5′ and 3′ end of the probe (forming the stem). When the probe is in a closed loop shape with the 5′ and 3′ ends hybridized to one another, the fluorescent reporter dye is quenched. When the molecular beacon probe hybridizes to the amplicon during PCR, the stem-and-loop structure opens, separating the fluorophore from the quencher releasing fluorescence. The application of molecular beacons in *Salmonella* detection in foods has been tested in a variety of food matrices (Bhaqwat et al., 2008; Patel & Bhagwat, 2008; Liming & Bhagwat, 2004).

Another variation of real-time PCR employs Scorpions technology. Scorpions are PCR primers covalently linked to a probe (Carters, et al., 2008). The reporter dye on the probe is prevented from fluorescing by a quencher dye on a separate complementary oligo. Upon primer extension, a probe-binding sequence is created which allows the probe to bind intramolecularly and generate fluorescence.

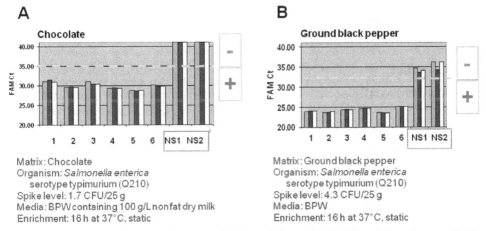

Matrix: Chocolate
Organism: *Salmonella enterica*
   serotype typimurium (Q210)
Spike level: 1.7 CFU/25 g
Media: BPW containing 100 g/L nonfat dry milk
Enrichment: 16 h at 37°C, static

Matrix: Ground black pepper
Organism: *Salmonella enterica*
   serotype typimurium (Q210)
Spike level: 4.3 CFU/25 g
Media: BPW
Enrichment: 16 h at 37°C, static

Fig. 4. Real-time PCR detection of *Salmonella*. 25 g of chocolate (A) 25 g of chocolate or (B) 25 g of black paper spiked with *Salmonella* were enriched for 16 hours. Six samples were spiked with 1-5 CFU *Salmonella* and 2 samples were not spiked with *Salmonella* (NS1 and NS2). Samples were prepared by the PrepSEQ® Rapid Spin Sample Preparation kit (in triplicate), and analyzed on the 7500Fast instrument using the MicroSEQ® *Salmonella spp.* Detection Kit.

RNA can also serve as the initial template for PCR for *Salmonella* detection. Detecting *Salmonella* RNA can serve as an indicator of viability of the bacteria (González-Escalona et al., 2009). RNA transcripts are likely present in higher copies than genomic DNA which can increase the sensitivity of the assay. In reverse transcription-PCR (RT-PCR), RNA is first converted to DNA by reverse transcriptase, and then PCR amplification occurs using the newly created DNA strands.

Alternative methods to real-time PCR have also recently emerged. The principle difference of these methods from PCR is that they use different approaches for generating new DNA or RNA with each cycle of amplification. In standard PCR, a denaturation step creates new DNA strands for DNA amplification to occur, theoretically doubling DNA template with each cycle. In contrast, loop-mediated isothermal amplification (LAMP) technology uses auto-cycling strand displacement DNA synthesis to create new DNA template. A *Bst* DNA polymerase large fragment with high strand displacement activity is added to the reaction. There are two general stages of LAMP: generation of template from the input sample and cycling amplification (Tomita et al., 2008). Typically, four primer sets are used to target six independent sequences flanking the target DNA. An inner primer hybridizes to the target DNA and elongates. This is followed by strand displacement which is primed by an outer primer; in this step, the outer primer hybridizes to the target DNA, displacing the newly created single strand DNA. Because each inner primer consists of a 5' overhang that becomes self-complementary to a sequence as the primer extends, the newly created single strand DNA forms a structure that has loops at each end. This DNA with stem-loop structure is the template of LAMP cycling. During LAMP cycling, the inner primer initiates auto-strand displacement with the template; also, self-priming occurs within the template. In short, the products are multiple stem-loop structures and elongated products containing the target sequence. LAMP reaction occurs at a constant temperature, typically 60-65 °C, and can be carried out in a water bath or heat block. Because a tremendous amount of DNA is formed in the reaction, the reaction by-product magnesium pyrophosphate forms a precipitate. The turbidity can be visible to the naked eye or visualized by UV after the addition of a fluorescent dye. The DNA products from LAMP can also be detected using a real-time turbidimeter. Application of LAMP in detection of *Salmonella* species in foods has previously been demonstrated (Ueda & Kuwabara, 2009). Another alternative method to real-time PCR is nucleic acid sequence-based amplification (NASBA) which amplifies RNA and creates new RNA strands by addition of a promoter site to complementary DNA. The isothermal method, which is typically run at 41 °C, first converts RNA into DNA using reverse transcriptase, then hydrolyzes RNA from the RNA-DNA hybrid using Rnase H. A target-specific primer, with a T7 promoter sequence at its 5' end, hybridizes to the single-stranded DNA. T7 RNA polymerase binds to the promoter region of the newly created double-stranded DNA to synthesize new RNA templates. Detection of the RNA product from NASBA can be accomplished by DNA hybridization followed by electrochemiluminescence. NASBA has previously been used to detect *Salmonella* Enteritidis in foods (D'Souza & Jaykus, 2003). Although there is no AOAC certified kit specifically for *Salmonella* detection by LAMP or NASBA, there are commercially available kits to perform these types of detection.

DNA hybridization is another useful way to detect target sequences of *Salmonella*. Labeled single-stranded DNA probes are added to a sample to detect either *Salmonella* DNA or RNA sequence. In one form of DNA hybridization, a poly-dA capture probe is added to a lysed sample in a microwell coated with poly-dT to detect *Salmonella*-specific rRNA sequence; simultaneously, a detector probe with a 5' horse-radish peroxidase (HRP) label is added to detect the same rRNA target (Mozola et al., 2007). Unhybrized probes are washed away, and after addition of a HRP substrate, the hybridization is detected by chemiluminescence. DNA hybridization can also be done using a dot blot format whereby a labeled probe is immobilized on a membrane (Iida et al., 1993).

Genetic technologies such as whole genome sequencing, microarrays, and SNP analysis are useful for identifying and typing *Salmonella*. However, these methods are not yet widely used for routine screening of food samples because they typically require more detailed workflows.

5. Analysis of real-time PCR data

In TaqMan® real-time PCR assay, the fluorescence released by the reporter dye during each cycle increases exponentially until the reaction reaches saturation. The cycle number at which the fluorescent signal first crosses the threshold value is the cycle threshold (C_T). It is dependent on the baseline which is established in the early cycles of the PCR. The threshold value, referred to as delta Rn (normalized reporter signal), can be adjusted by the user. The threshold value should be set above baseline noise in the early cycles of PCR and within the exponential amplification phase for positive samples (Figure 5). Using PCR replicates of positive samples is a useful way to decide upon the appropriate threshold to use for data analysis. The appropriate delta Rn value will be dependent upon the assay and application and, once set, should be used for all samples for consistent analysis. Low C_T values indicate high copy number of the target sequence, while high C_T values indicate low copy number. The typical limit of detection is 10 to 100 copies of purified genomic DNA and 10^3 to 10^4 CFU/mL before sample preparation. Most assays for *Salmonella* detection are typically qualitative, producing data that can be categorized as positive or negative. However, assay runs can be designed to be quantitative. Quantitative real-time assays should be validated and shown to amplify with high efficiency. The linear range for quantitation should be based on the exponential phase of amplification. Unknown samples to be quantitated should be run alongside DNA standards during real-time PCR.

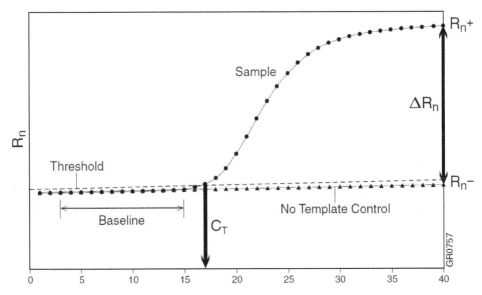

Fig. 5. Demonstration of a TaqMan® amplification curve and parameters. Amplification results in increased fluorescence (Rn). The Ct value reflects the beginning of amplification.

In PCR using SYBR® Green dye, a real-time amplification plot is generated that is similar to a TaqMan® PCR assay. Melt curve analysis (also known as a dissociation curve analysis) is typically added to the end of a SYBR® Green PCR. During the dissociation stage, the instrument increases in temperature over several minutes. For positive samples, SYBR® Green is initially bound to the amplicons. As the double-stranded amplicon dissociates, there is a drop in SYBR® Green fluorescence. The change in fluorescence is plotted against the temperature (Figure 6). Unknown samples should be compared to no-template reactions. Samples that are positive should have product of the expected Tm and, if the assay is well-designed, should not have other products such as primer-dimer or mis-primed products.

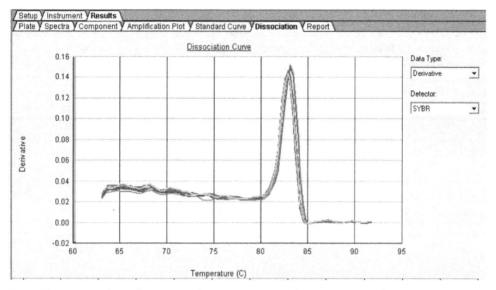

Fig. 6. Derivative of a melting curve from a reaction with SYBR® Green dye reaction. The Tm is determined by the peak of the derivative.

6. Software

Easy-to-use software is an important component of applying genetic methods to detect Salmonella. It is crucial to have accurate, reproducible, and unbiased data interpretation. Results should be clearly stated in convenient formats to the end user. In RapidFinder™ Express Software (Life Technologies) for food pathogen detection, algorithms were developed to interpret real-time PCR data and allow for sensitive detection while avoiding false positive results. Results are stated as positive or negative depending on cutoff values. In cases in which a positive or negative assessment cannot be made (e.g, the internal positive control failed indicating inhibition), a warning assessment is generated along with an explanation (Figure 7).

7. Bioinformatic tools for real-time PCR assays design

Selecting a good genomic target is critical to designing a real-time PCR assay. Appropriate genomic targets have to be sequenced from multiple strains in order to design highly specific primers and probes which cover a set of desired target species (*inclusion set*) and exclude from detecting a set of other bacterial genomes (*exclusion set*). Examples of broadly sequenced bacterial DNA targets are the 16S gene, *gapA*, *recA*, *rpoB*, among others. We developed and validated a standard bioinformatics assay design tool to generate primer and probe combinations for real-time PCR pathogen detection. The first step of the design is generation of a target consensus sequence based on multiple sequence alignments of all available target sequences using clustalW algorithm (Larkin et al., 2007). Sequences can be used from all available public databases. At the second step, a set of well described selection criteria (Larkin et al., 2007; Endrizzi et al., 2002; Kramer & Coen, 2000) has to be applied, such as optimal Tm, nucleotide distribution (*e.g.* avoid high GC content and poly-N stretches), absence of cross-hybridization; and amplicon size (optimal size is 60-150 base pairs). This set of rules can be applied to select candidate primer and probe sequences that target a signature nucleic acid sequence in a microbe of interest. A set of optimal assays has to be evaluated, considering criteria of hybridization patterns of the two primers and a TaqMan® probe to the intended target sequence (Endrizzi et al., 2002). At the third step, those assays with the highest specificity are selected. The specificity is determined based on nucleic acid sequence comparisons of the binding sites for the assay primers and probes with genomic sequences from other bacterial species (exclusion set), being sure to include closely related species in the analysis. These genomic sequences have to be from multiple available databases, as well as from additional sequences that can be specifically generated for the design of the TaqMan® detection assays. Based on this sequence comparison, specific primers and probes can be selected. The best primers contain the highest number of mismatches to other non-target bacteria genomes, with mismatches preferentially located at the 3'-prime region of the primers. This minimizes the possibility that an assay will be selected that generates a false-positive signal (Furtado et al., 2004).

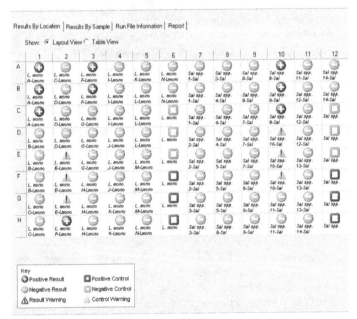

Fig. 7. RapidFinder™ Express Software for food pathogen detection displays positive, negative, and warning assessments for real-time PCR.

Additional DNA targets can be amplified and sequenced. This is often required when inclusion/exclusion testing show non-detection of inclusion strains or unwanted detection of exclusion strains. This requires the design of sequencing primers upstream and downstream of the assay target. It might also be useful to create a specific signature sequence database by sequencing additional samples, which can increase the confidence of specific assay designs. Sequence files can be analyzed using clustalW multiple sequence alignment. The Applied Biosystems® MicroSEQ® ID validated method can be used for sequencing the 16S fragment. Using the 16S gene as a target for specific pathogen detection is challenging because it contains conservative sequences across multiple bacterial species. However, when primers and probes have to detect a broad range of species, families, genus or higher order, the 16S target is ideal or often is the only choice.

Experimental validation of the assay specificity is a critical step in the final assay selection. To test for assay specificity, a panel containing a diverse group of *bacteria* isolates (target and exclusion strains) has to be established and used for experimental validation. The panel must be well characterized by stereotyping or sequence-based bacterial identification (Tanner et al., 2006). Preferably, the panel will contain ATCC type-strains.

8. *Salmonella* detection background

More than 2,400 *Salmonella* serotypes have been reported, all of which are potentially pathogenic. The species *Salmonella* enterica, with its 6 main subspecies is of clinical relevance for humans and is the causative agent of food borne illnesses or salmonellosis. However, for tracking purposes it is often important to have a very specific serotype assay to measure and contain the spread of an outbreak pathogen. Food borne outbreaks due to *Salmonella* have become a major public health problem and can occur either as food poisoning triggered epidemics or as isolated cases. Outbreaks have been associated with raw meats and poultry, eggs, milk and dairy products, seafood, coconut sauces, salad dressings, cocoa, chocolate, and peanuts. The combination of efficient sample preparation protocols and reliable detection of *Salmonella* is a solid path forward for molecular detection methods.

9. Design options: Single-plex, multiplex, degenerate primers and probes & optimization

The most important advantages of real-time PCR are the ability for quantification of target concentration and multiplexing with different dyes per target. Since real-time PCR measures target amplification by monitoring the increase in fluorescence generated by probe degradation, the only limitation to multiplexing within a single reaction tube is the number of fluorophores that can be distinguished by the detection system (i.e. optics and software). From the chemistry side, one of the keys to maximizing the number of fluorogenic signals measured in a single reaction mix is developing dyes with well-separated and narrow emission spectra. Life Technologies identified several dye sets and real-time PCR systems that allow for detection of up to five fluorophores in a single reaction mix. It is important to note that instruments have different detection ranges for monitoring fluorescent energy. Detecting multiple targets can also be achieved through novel engineering designs; an example being the 384-well format Custom TaqMan® Array card that can be used in combination with the Applied Biosystems 7900HT Fast real-time PCR System (Tebbs et al., 2010), and the OpenArray® real-time PCR System. These Systems split a common sample amongst multiple reaction chambers prior to real-time PCR amplification and detection.

Since the 7500Fast real-time PCR system capability allows for measuring five fluorophores in a single reaction mix, it is possible to create real-time reaction applications capable of detecting 5 target organisms in a single tube. However, for high accuracy applications, two of the five channels are often used as system controls. We created a number of applications where one channel is assigned to the internal positive control (IPC) to monitor the presence of PCR inhibitors, and another channel is used to control for well-to-well variation and normalization of fluorescence detection by using a ROX™ dye as a standard dye. Detection of IPC is indicative of a successful PCR amplification. Thus, with the inclusion of two controls, three targets can be detected in a single reaction tube. Each assay target is evaluated independently through the dye assigned to that target. Important considerations when designing multiplex real-time PCR assays include: first, development of several specific working assays for each target to have a choice in a final configuration. Each assay should be tested for quantification efficiency. Second, each assay must be tested against a large panel of microorganisms that include both inclusive and exclusive strains to identify highly specific assays. It is possible to add additional primers and probes containing variant base sequences (degenerate sequences) if some inclusive strains are not detected or weakly detected. Third, fluorogenic dye signals should be balanced by adjusting primer/probe concentrations: it is essential to perform statistics-based DOE (Design of Experiments) studies to optimize primer and probe concentrations for optimal detection of each target species. The final multiplex assay can be further optimized by standard PCR optimization techniques including adjusting magnesium and enzyme concentration, annealing temperature, probe lengths, and instrument settings.

10. Design and validation of existing real-time PCR *Salmonella* assays

Specific assays were designed and tested for detection of (i) *Salmonella* species in food samples, (ii) *Salmonella* Enteriditis in eggs and environmental samples, and (iii) *Salmonella* Typhimurium (Table 3). The selection of the target genes was based on specific applications: whether targeting detection of all *Salmonella* species in one reaction or targeting a specific serotype. Experimental validation is an essential part of molecular assay development. Testing and validation of the complete workflow is a critical element in the acceptance of a detection assay in food safety testing laboratories.

Salmonella **target organism**	**Gene Target**
Salmonella species	*hilA* gene
Salmonella Enteriditis	*Prot6e* gene
Salmonella Typhimurium	*Target 1/Target 2*

Table 3. Real-time PCR assays for *Salmonella*.

Our *Salmonella* spp. detection assay is a rapid, sensitive real-time PCR test, that, when combined with the automated PrepSEQ® Nucleic Acid extraction method or the manual PrepSEQ® Rapid Spin extraction method, allows the completion of *Salmonella* spp. detection within 18-19 hours, compared to 3-5 days required by for the traditional culture-based methods (Andrews & Hammack, 2011). As described in the previous section, each real-time PCR reaction contains an Internal Positive Control (IPC) that monitors for the presence of inhibitors for reliable negative results. The assay was designed as a complete reaction mix (no target DNA) in a lyophilized format to allow for minimal pipetting steps and addition of maximum sample volume. This workflow allows for robust *Salmonella* detection and creates the possibility of testing composite or pooled samples reliably.

The assay was validated for *Salmonella* detection in different food matrices including a variety of food matrices which were previously associated with food recalls or outbreaks (Carroll, 2009; Cahill et al., 2008; van Cauteren et al., 2009; Munnoch et al., 2009; Reiter et al., 2007): raw ground beef, raw chicken wings, chocolate, raw shrimp, Brie cheese, shell eggs, cantaloupe, black pepper, dry infant formula, and dry pet food. The detection was complemented by two sample preparation protocols for flexible detection work-flow set up: manual Rapid Spin column-based method and high-throughput automated method. The complete workflow was evaluated against the reference ISO 6579 culture confirmation method. Examples for detecting *Salmonella* in "difficult" matrices chocolate and black pepper are shown in Figure 4 (above). The sensitivity and specificity rates for *Salmonella* spp. detection assays were 100%, with no false negative or false positive samples observed with both sample preparation methods. Inclusivity panel contained 100 *Salmonella* strains based on variety of serotypes. Exclusivity panel contained 30 different bacterial isolates, including genetically close microorganisms as well as bacteria that are common organisms in the environment (Balachandran et al., 2011).

A special study was conducted to demonstrate detection of *Salmonella* in peanut butter, as a part of the Emergency Response Validation program, that followed the peanut butter outbreak in the United States in 2009 (Tebbs et al., 2009). The method was evaluated using *S. enterica* ser. Typhimurium ATCC14028 strain and the reference FDA-BAM protocol was used as culture confirmation method. There was complete agreement between the automated PrepSEQ® NA Extraction method and culture confirmation for uninoculated samples (see Table 5 in Tebbs et al., 2009). Also, Chi-square analyses indicated that the proportions of positives for the *Salmonella* spp. detection method and the reference method were not statistically different at the 5% level of significance.

A specific *Salmonella* Typhimurium assay was designed using duplex assay approach: two different assay probes carry different fluorescent dyes. This two-target approach increases confidence calling the test positive: both fluorescent signals should be positive (Figure 8). An example of positive *Salmonella* Typhimurium detection using this two-target assay is presented in Figure 9.

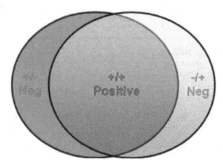

Fig. 8. Diagram representing a duplex real-time PCR assay for detection of *Salmonella* Typhimurium. Each assay detects *S.* Typhimurium plus some non-Typhimurium strains. Because each assay detects a different set of non-Typhimurium strains, only when both assays are positive is the sample positive for *S.* Typhimurium. The assays are labeled with a different fluorophore to be detected independently by real-time PCR.

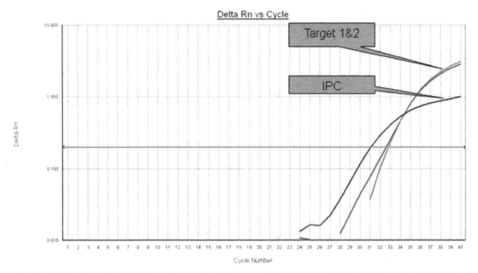

Fig. 9. Example showing positive detection of *Salmonella* Typhimurium with a two-target assay. The assay for target 1 uses FAM™, the assay for target 2 uses VIC®, and the assay for IPC uses NED™. The assay was analyzed on the 7500Fast real-time PCR instrument.

Real-time PCR methods for detecting foodborne pathogens offer the advantages of simplicity and quick time to results compared to traditional culture methods. Our assays demonstrated high accuracy detection of *Salmonella* strains in inclusivity panels, and good discrimination against detection of exclusivity panels.

11. Conclusion

Rapid methods offer great advantages to food producers minimizing risks associated with long hold times during pathogen testing. This is clearly illustrated with requirements for testing whole shell eggs for *Salmonella enterica* serovar Enteritidis (SE). In July of 2009 the U.S. Food and Drug Administration announced The Federal Egg Safety Program, a new regulation that requires routine environmental tests of poultry houses for presence of *Salmonella* Enteritidis (Food and Drug Administration, 2009). If SE is present in the environment, there is a requirement to test eggs prior to their distribution for sale. According to the regulation, 50 egg pools consisting of 20 eggs per pool must be tested every 2 weeks for 8 weeks. The traditional method, designed and approved by the FDA, takes up to ten days to get results (Andrews & Hammack, 2011). According to the FDA approved method, egg pools sit for 4 days at room temperature to allow growth of SE, and then a 25 gram sample is pre-enriched in modified typtic soy broth with ferrous sulfate. The sample is then grown in selective media, then selective agars for presumptive detection, and then confirmed by biochemical and serological methods. The procedure is laborious and expensive. A recently developed real-time PCR assay allows detection of SE in egg pools in less than 27 hours. The TaqMan® *Salmonella* Enteritidis Detection Kit (Life Technologies) enriches egg pool samples for 24 hours, following which the samples are prepped and combined with a PCR reaction mix for detection by real-time PCR. Because sample prep is fully automated, the total hands on time following enrichment is less than one hour. A method comparison study showed that the real-time PCR method was equivalent to the FDA reference method (Table 4). The FDA reported that the real-time PCR kit was equivalent to the FDA BAM Chapter 5, *Salmonella* method for detection of *Salmonella* Enteritidis in accuracy, precision, and sensitivity (FDA website). The Pennsylvania Layer Industry and Penn Ag approved the use of the real-time PCR method as an option for testing of egg and environmental samples without the need for culture confirmation. The trend appears to show increased acceptance of real-time PCR and other fast methods as alternatives to the more cumbersome culture methods.

Inoculation Level	Inoculating Organism	U.S. FDA BAM	TaqMan® Salmonella Enteritidis Method		x^2	Relative Sensitivity	False Negative Rate	False Positive Rate
			Presumed	Confirmed				
Experiment 1								
Control	N/A	0/5	0/5	0/5	-	-	0%	0%
Spike	S. enterica ser. Enteritidis ATCC 13076	16/20	16/20	16/20	0	100%	0%	0%
Experiment 2								
Control	N/A	0/5	0/5	0/5	-	-	0%	0%
Spike	S. enterica ser. Enteritidis ATCC 13076	11/20	13/20	13/20	0.41	118%	0%	0%

Table 4. Methods comparison showed that the TaqMan® Real-time PCR method was equivalent to the FDA BAM method for detection of *Salmonella* Enteritidis in whole shell eggs. The results from chi-square analysis on two independent experiments indicated no difference between the two methods (χ^2 = 0 and 0.41 for Experiment 1 and Experiment 2, respectively). No false positive or false negative results were observed.

12. References

Al-Soud WA, Radstrom P (2001). Purification and characterization of PCR-inhibitory components in blood cells. *Journal of Clinical Microbiology*, 39(2):485-493.

Andersen MR, Omiecinski CJ (1992). Direct extraction of bacterial plasmids from food for polymerase chain reaction amplification. *Applied and Environmental Microbiology* 58(12):4080-4082.

Andrews WH, HammackT (2011). Bacteriological Analytical Manual, Chapter 5, *Salmonella*. In *U.S. Food and Drug Administration, Bacteriological Analytical Manual Online* (http://www.fda.gov/Food/ScienceResearch/LaboratoryMethods/Bacteriological AnalyticalManualBAM/ucm070149.htm). Accessed July 27, 2011.

AOAC Research Institute website. (http://www.aoac.org/testkits/testkits.html). Accessed July 27, 2011.

Atmar RL, Metcalf TG, Neill FH, Estes MK (1993). Detection of enteric viruses in oysters by using the polymerase chain reaction. *Applied and Environmental Microbiology* 59(2):631-635.

Balachandran P, Cao Y, Wong L, Furtado MR, Petrauskene OV, Tebbs RS (2011). Evaluation of Applied Biosystems MicroSEQ® Real-Time PCR system for detecgtion of *Salmonella* spp. in food: *Performance Tested Method*SM 031001. *Journal of AOAC International*, 94(4):1106-1116.

Bhagwat AA, Patel J, Chua T, Chan A, Cruz SR, Aguilar GA (2008). Detection of *Salmonella* species in foodstuffs. *Methods in Molecular Biology*, 429(II):33-43.

Bolsilevac JM, Kalchayanand N, Schmidt JW, Shackelford SD, Wheeler TL, Koohmaraie M (2010). Inoculation of beef with low concentrations of Escherichia coli O157:H7 and examination of factors that interfere with its detection by culture isolation and rapid methods. *Journal of Food Protection*, 73(12):2180-2188.

Boyd EF, Wang FS, Whittam TS, Selander RK (1996). Molecular genetic relationships of the *salmonellae*. *Applied and Environmental Microbiology*, 62(3):804–808.

Cahill SM, Wachsmuth IK, Costarrica Mde L, et al (2008). Powdered infant formula as a source of *Salmonella* infection in infants. *Clin Infect Dis.*, 46:268-273.

Carroll C (2009). Corporate Reputation Review 12, 64-82.

Carters R, Ferguson J, Gaut R, Ravetto P, Thelwell N, Whitcombe D (2008). Design and use of scorpions fluorescent signaling molecules. *Methods in Molecular Biology*, 429(II):99-115.

Center for Disease Control (2004). *Salmonella* Annual Summary 2004.

Coenye T, Vandamme P (2003).Extracting phylogenetic information from whole-genome sequencing projects: the lactic acid bacteria as a test case. *Microbiology*, 149(Pt 12):3507–3517

Demeke T, Adams RP (1992). The effects of plant polysaccharides and buffer additives on PCR. *BioTechniques*, 12(3):332-334.

Department of Agriculture (2011). Not applying the mark of inspection pending certain test results. In *Federal Register*, 76(69):19952-19970.

D'Souza, D.H., Jaykus, L.-A. (2003). Nucleic acid sequence based amplification for the rapid and sensitive detection of *Salmonella enterica* from foods. *J Applied Microbiol.* 95, 1343–1350.

Eckhart L, Bach J, Ban J, Tschachler E (2000). Melanin binds reversibly to thermostable DNA polymerase and inhibits its activity. *Biochemical and Biophysical Research Communications*, 271(3):726-730.

Endrizzi K, Fischer J, Klein K, Schwab M, Nussler A, Neuhaus P, Eichelbaum M, Zanger UM (2002). Discriminative quantification of cytochrome P4502D6 and 2D7/8 pseudogene expression by TaqMan real-time reverse transcriptase polymerase chain reaction. *Anal. Biochem.*, 300: 121-131.

Fitzmaurice J (2006). Detection of verotoxin genes VT 1 and VT 2 in Escherichia coli O157:H7 in minced beef using immunocapture and real-time PCR. *Methods in Molecular Biology*, 345:91-96.

Fleischmann RD, Adams MD, White O, Clayton RA, Kirkness EF, Kerlavage AR, Bult CJ, Tomb JF, Dougherty BA, Merrick JM, et al. (1995). Whole-genome random sequencing and assembly of *Haemophilus influenza*. Rd. *Science*, 269(5223):496-512.

Food and Drug Administration website, Testing methodology for *Salmonella* Enteritidis (SE), (http://www.fda.gov/Food/FoodSafety/Product-SpecificInformation/EggSafety/EggSafetyActionPlan/ucm228796.htm). Accessed July 26, 2011.

Food and Drug Administration (2008). Environmental sampling and detection of *Salmonella* in poultry houses. In *U. S. Food and Drug Administration website* (http://www.fda.gov/Food/ScienceResearch/LaboratoryMethods/ucm114716.htm). Accessed July 27, 2011.

Food and Drug Administration (2009). Prevention of *Salmonella* Enteriditis in Shell Eggs During Production, Storage, and Transportation; Final Rule (July 9, 2009). *Federal Register*, Vol 74 (130): 33030-33101.

Fratamico P, Bagi LK, Cray WC, Narang N, Yan X, Median M, Liu M (2011). Detection by multiplex real-time polymerase chain reaction assays and isolation of shiga toxin-producing *Escherichia coli* serogroups 026, 045, 0103, 0111, 0121, and 0145 in ground beef. *Foodborne Pathogens and Disease*, 8(5):601-607.

Furtado M.R, Petrauskene OV, Livak KJ (2004). Application of Real-Time Quantitative PCR in the Analysis of Gene Expression. In *Horizon Bioscience*. DNA Amplification: Current Technologies and Applications. 131-145.

González-Escalona N, Hammack TS, Russell M, Jacobson AP, De Jesús AJ, Brown EW, Lampel KA (2009). Detection of live *Salmonella* spp. cells in produce by a TaqMan-based quantitative reverse transcriptase real-time PCR targeting invA mRNA. *Applied and Environmental Microbiology*, 75(11):3714-3720.

Healy B, Huynh S, Mullane N, O'Brien S, Iversen C, Lehner A, Stephan R , Parker CT, Fanning S (2009). Microarray-based comparative genomic indexing of the *Cronobacter* genus (*Enterobacter sakazakii*). *International Journal of Food Microbiology*, 136(2):159-164.

Iida K, Abe A, Matsui H, Danbara H, Wakayama S, Kawahara K (1993). Rapid and sensitive method for detection of *Salmonella* strains using a combination of polymerase chain reaction and reverse dot-blot hybridization. *FEMS Microbiol Lett.*, 114(2):167-172.

ISO 6579:2002 (E). *Microbiology of food and animal feeding stuffs - Horizontal method for the detection of Salmonella spp.* (2002) 4th Ed., INTERNATIONAL STANDARD, Geneva, Switzerlan

Iversen C, Mullane N, McCardell B, Tall BD, Lehner A, Fanning S, Stephan R, Joosten H (2008). *Cronobacter* gen. nov., a new genus to accommodate the biogroups of *Enterobacter sakazakii*, and proposal of *Cronobacter sakazakii* gen. nov., comb. nov., *Cronobacter malonaticus* sp. nov., *Cronobacter turicensis* sp. nov., *Cronobacter muytjensii* sp. nov., *Cronobacter dublinensis* sp. nov., *Cronobacter* genomospecies 1, and of three subspecies, *Cronobacter dublinensis* subsp. dublinensis subsp. nov., *Cronobacter dublinensis* subsp. lausannensis subsp. nov. and *Cronobacter dublinensis* subsp. lactaridi subsp. nov. *International Journal of Systematic and Evolutionary Microbiology*, 58(Pt 6):1442-1447.

Kramer MF, and Coen DM (2000). Enzymatic Amplification of DNA by PCR: Standard Procedures and Optimization. In *Current Protocols in Toxicology*. A.3C.1-A.3C.14.

Kunin V, Ahren D, Goldovsky L , Janssen P, Ouzounis CA (2005). Measuring genome conservation across taxa: divided strains and united kingdoms. *Nucleic Acids Research*, 33(2):616-621.

Larkin MA, Blackshields G, Brown NP, Chenna R, McGettigan PA, McWilliam H, Valentin F, Wallace IM, Wilm A, Lopez R, Thompson JD, Gibson TJ, Higgins DG (2007). ClustalW and ClustalX version 2. *Bioinformatics*, 23: 2947-2948.

Liming SH, Bhagwat AA (2004). Application of a molecular beacon-real-time PCR technology to detect *Salmonella* species contaminating fruits and vegetables. *International Journal of Food Microbiology*, 95(2):177-187.

Lofstrom C, Knutsson R, AAxelsson CE, Radstrom P (2004). Rapid and specific detection of *Salmonella* spp. in animal feed samples by PCR and culture enrichment. *Applied and Environmental Microbiology,* 70(1):69-75.

McGregor DP, Forster S, Steven J, Adair J, Leary SE, Leslie DL, Harris WJ, Titball RW (1996). Simultaneous detection of microorganisms in soil suspension based on PCR amplification of bacterial 16S rRNA fragments. *Biotechniques,* 21(3):463-6, 468, 470-1.

McQuiston JR, Herrera-Leon S, Wertheim BC, Doyle J, Fields PI, Tauxe RV, Logsdon JM (2008). Molecular phylogeny of the salmonellae: relationships among *Salmonella* species and subspecies determined from four housekeeping genes and evidence of lateral gene transfer events. *Journal of Bacteriology,* 190(21):7060-1067.

Monteiro L, Bonnesaison D, Vekris A, Petry KG, Bonnet J, Vidal R, Cabrita J, Megraud F (1997). Complex polysaccharides as PCR inhibitors in feces: *Helicobacter pylori* model. *Journal of Clinical Microbiology,* 35(4):995-998.

Mozola MA, Peng X, Wendorf M (2007). Evaluation of the GeneQuence DNA hybridization method in conjunction with 24-hour enrichment protocols for detection of *Salmonella* spp. in select foods: collaborative study. *Journal of AOAC International,* 90(3):738-755.

Munnoch SA, Ward K, Sheridan S, Fitzsimmons GJ, Shadbolt CT, Piispanen JP, Wang Q, Ward TJ, Worgan TL, Oxenford C, Musto JA, McAnulty J, Durrheim DN (2009). A multi-state outbreak of *Salmonella* Saintpaul in Australia associated with cantaloupe consumption. *Epidemiology and Infection,* 137(3):367-374.

Nam HM, Srinivasan V, Gillespie BE, Murinda SE, Oliver SP (2005). Application of SYBR green real-time PCR assay for specific detection of *Salmonella* spp. in dairy farm environmental samples. *International Journal of Food Microbiology,* 102(2):161-71.

Patel JR, Bhagwat AA (2008). Rapid real-time PCR assay for detecting *Salmonella* in raw and ready-to-eat meats. *Acta Vet Hung.,* 56(4):451-8.

Porwollik S, Wong RM McClelland M (2002). Evolutionary genomics of *Salmonella*: gene acquisitions revealed by microarray analysis. *Proceedings of the National Academy of Sciences,* 99(13):8956-8961.

Pupo GM, Lan R, Reeves PR (2000). Multiple independent origins of *Shigella* clones of *Escherichia coli* and convergent evolution of many of their characteristics. *Proceedings of the National Academy of Sciences,* 97:(19)10567-10572.

Reiter MG, Fiorese ML, Moretto G, López MC, Jordano R (2007). Prevalence of *Salmonella* in poultry slaughterhouse. *J. Food Prot.,* 70:1723-1725.

Rossen L, Norskov P, Holmstrom K, Rasmussen OF (1992). Inhibition of PCR by components of food samles, microbial diagnostic assays and DNA-extraction solutions. *International Journal of Food Protection,* 17:37-45.

Tanner MA, Scarpati E, Sharma S, Furtado MR (2006). MicroSeq®: A 16S rRNA Gene Sequencing System for Bacterial Identification. In Volume III of Encyclopedia of Rapid Microbiological Methods. 205-225.

Tebbs RS, Cao YY, Balachandran P, Petrauskene O (2009). TaqMan *Salmonella* enterica Detection Kit. Performance Tested Method 020803. *Journal of AOAC International,* 92(6):1895-901.

Tebbs RS, Liu J-K, Furtado MR, Petrauskene O (2010). Mulitiplex detection of pathogenic organisms using TaqMan real-time PCR. In *Biodetection Technologies, Technical Responses to Biological Threats,* 6th Edition, Knowledge Press, pages 5-13.

Techathuvanan C, D'Souza DH (2011). Optimization of rapid *Salmonella* enterica detection in liquid whole eggs by SYBR green I-Based real-time reverse transcriptase-polymerase chain reaction. *Foodborne Pathogens and Disease*, 8(4):527-534.

Tindall BJ, Grimont PA, Garrity GM, Euzeby JP (2005). Nomenclature and taxonomy of the genus *Salmonella*. *International Journal of Systematic and Evolutionary Microbiology*, 55(Pt 1):521-524.

Tomita N, Mori Y, Kanda H, Notomi T (2008). Loop-mediated isothermal amplification (LAMP) of gene sequences and simple visual detection of products. *Nat Protoc.* 3(5):877-882.

Tsai Y-L, Olson BH (1992a). Detection of low numbers of bacterial cells in soils and sediments by polymerase chain reaction. *Applied and Environmental Microbiology*, 58(2):754-757.

Tsai Y-L, Olson BH (1992b). Rapid method for separation of bacterial DNA from humic substances in sediments for polymerase chain reaction. *Applied and Environmental Microbiology*, 58(7):2292-2295.

Ueda, S. and Kuwabara, Y. (2009). The rapid detection of *Salmonella* from food samples by loop-mediated isothermal amplification. *Biocontrol Science.* 14(2), 73-76.

van Cauteren D, Jourdan-da Silva N, Weill FX, King L, Brisabois A, Delmas G, Vaillant V, de Valk H (2009). Outbreak of *Salmonella enterica* serotype Muenster infections associated with goat's cheese, France, March 2008. *Eurosurveillance*, 14(31):19290.

Weyant RS, Edmonds P, Swaminathan B (1990). Effect of ionic and nonionic detergents on the *Taq* polymerase. *BioTechniques*, 9(3)308-309.

Wilson IG (1997). Inhibition and facilitation of nucleic acid amplification. *Applied and Environmental Microbiology*, 63(10):3741-3751.

5

A Tale of 6 Sigmas: How Changing Partners Allows *Salmonella* to Thrive in the Best of Times and Survive the Worst of Times

R. Margaret Wallen and Michael H. Perlin

University of Louisville
USA

1. Introduction

Salmonella enterica are rod-shaped, facultative anaerobic, Gram-negative members of the Enterobacteriacae family (Dougan et al., 2011). Most people have heard of the bacteria and generally associate it with food-borne illness. Despite general public knowledge of the health risks associated with and precautions taken to prevent its spread, *Salmonella* continues to cause many problems. One approach toward curbing this spread and reducing the negative impact of *S. enterica* could be genetic analysis, with an ultimate goal of understanding why the bacteria are able to survive attempts to destroy them.

It has been suggested that the *Salmonella* genus diverged from *Escherichia coli* somewhere between 100 and 150 million years ago (Dougan et al., 2011). While there is evolutionary distance between the two genera, much of the genetic information has been conserved, and as a result, the study of one organism has provided insight into the study of the other. *Salmonella* spp. are generally considered to be pathogenic and can have both warm- and cold-blooded hosts (Dougan et al., 2011). More recent evolution has occurred within the *Salmonella* genus itself. *Salmonella enterica* has evolved into many different subspecies and serovars that manifest in dramatically different ways across a variety of hosts despite sharing 95% of the same genetic information (McDermott et al., 2011). From a medical perspective, *Salmonella* genetics are particularly important. Although a single-celled organism, due to its long evolutionary history with humans and other organisms, these bacteria has developed several sophisticated mechanisms to survive the immune systems of its hosts and evade sanitation efforts to kill it. Understanding how this survival at the most fundamental of levels, it may be possible to more specifically combat the bacteria.

Salmonella typically reach their hosts through the consumption of contaminated food or water. Once inside its host, the bacteria must persist through various levels of pH, temperature, osmolarity, and nutrient availability (Ohl & Miller, 2001). The pathogen must also face various attempts by the host's immune system to eliminate it. Each different environment and each assault on the bacteria's integrity must be addressed by the organism in order to survive. The ability of the organism to thrive in a multitude of environments and persist to establish infection in its host is governed by the expression of different genes.

While there are a multitude of regulatory pathways within *Salmonella* that can influence gene expression, one of the most fundamental comes from the usage of alternate sigma factors by the cell's RNA polymerase, as is the case for most prokaryotes. Sigma factors facilitate differential gene expression by reversibly binding to the RNA polymerase core enzyme and providing specificity for certain promoter regions. The various sigma factors have different affinities for particular promoters as well as for the core enzyme itself. Similar to other cellular proteins, sigma factors are regulated at a variety of levels. Transcription in *Salmonella*, as in all prokaryotes, requires a sigma factor, and ultimately all gene expression is affected by sigma factors and their activity.

Sigma factors were originally discovered as protein factors that stimulate RNA synthesis from DNA using DNA-dependent RNA polymerase (Burgess & Travers, 1969). These proteins all share four regions of similarity indicative of a common function (Kutasake et al., 1994). For the group of closely related sigma factors, special regions within the protein recognize specific regions of the DNA as promoters versus non-promoter regions of DNA (Dombroski et al., 1992). These DNA regions include conserved sequences centered around the -35 and -10 positions with respect to the transcription initiation site. By truncating the sigma protein at various locations, researchers were able to determine that four conserved regions of the sigma factors were responsible for locating different areas of the promoter region. For example, region 4 of the sigma factor is found to recognize the consensus sequence around -35, while regions 2 and 3 recognize the -10 consensus sequence (Dombroski et al., 1992). Region 1 of the sigma factor, the amino terminus of the protein, blocks regions 2, 3, and 4 from interacting with the DNA (Dombroski et al., 1993). Binding of the sigma factor to the core enzyme blocks region 1 and allows interaction of the other three regions with the DNA (Dombroski et al., 1992). In this way, the sigma factor cannot interact with DNA without being bound by RNA polymerase. While it was understood that a sigma factor was necessary to facilitate transcription, their power to regulate gene expression was not fully understood.

2. Early virulence-related genetic studies

As with most pathogenic microorganisms, early genetic research focused on the disease-causing properties of *Salmonella*. Preliminary studies involving virulence properties of *Samonella* revealed that in the absence of a functional copy of several genes, the bacteria was unable to survive to cause infection inside its host. Further studies of each of these genes revealed that while all of the genes were required for optimal virulence, the gene expression was not under the same regulatory control. Baumler and his colleagues examined nearly 30 mutant strains of *Salmonella* Typhimurium that had shown attenuated ability to infect and survive inside mouse macrophages (Baumler et al., 1994). Baumler concluded that these genes all made contributions to the virulence properties of *Salmonella*.

Some of the particular genes that Baumler concluded were disrupted in the attenuated strains were *purD*, *prc*, *fliD*, and *nagA* (Baumler et al., 1994). Other researchers have examined the transcriptional control of these genes to understand why they are so essential to the virulence capabilities of *Salmonella*.

As many sigma factors are closely related, there is a high degree of homology between their structures and therefore promoter affinities. However, as few as one or two base pair change can dramatically change which sigma factor recognizes the promoter (Römling et al., 1998). The *purD* gene encodes 5'-phosphoribosylglycinamide synthetase, which is involved in purine nucleotide synthesis (Aiba & Mizobuchi, 1989). While these genes have easily identifiable -10 consensus sequences, none appear to have the -35 region similar to those typically recognized by the primary sigma factor (Kilstrup et al., 1998). Potentially, the ambiguity of the promoter region shows the ability to be used by multiple sigma factors.

A second gene, *prc*, encodes a protease that in closely related organisms has been found to play a role in response to cell wall stress (Wood et al., 2006). In these organisms, *prc* is preceded by a consensus sequence for a sigma factor showing a great deal of similarity to the sigma factor in *E. coli* and *Salmonella* that responds to a variety of global stresses, including damage to the cellular envelope (Wood et al., 2006).

The *fliD* gene encodes part of the flagella filament, needed for the motility of the bacteria (Kutsukake et al., 1994). This gene is proceeded by a consensus sequence that can only be used by the flagella-specific sigma factor (Kutsukake et al., 1994) and is part of a highly temporally and spatially regulated pathway that ensures flagella are expressed readily in times that motility is necessary and repressed when the bacteria have not formed the appropriate primary structures for the flagellar.

The *nagA* gene product is N-acetylglucosamine-6-phosphate deacetylase in *E. coli* and has the same function in *Salmonella* Typhimurium (Baumler et al., 1994). This gene was found to have consensus sequence in the -10 region requiring the activation of a magnesium sensitive regulator in the presence of the housekeeping sigma (Minagawa et al., 2003). Based only on the extracellular availability of magnesium, the primary sigma factor is responsible for the transcription of the gene, provided a secondary regulatory system is activated.

With the genes that Baumler examined, in combination with other research indicating that each of these types of genes was under different regulatory control by particular sigma factors, a pattern began to emerge. Genes responsible for the organism's response to particular threats to its integrity were under the transcriptional direction of particular sigma factors. The importance of sigma factors as transcriptional regulators is further revealed by their stability over time (Sutton et al., 2000) and the high degree of homology between closely related species (Guiney et al., 1995).

3. A tale of six sigmas

To date, six different sigma factors have been discovered to be encoded within the *Salmonella* genome that are responsible for transcription from a variety of promoters in response to different phases of the organism's life as well as environmental conditions. Acting together in a complex, as an interconnected web of gene regulation, they enable *Salmonella* to withstand and thrive inside infected hosts.

Sigma factors were characterized as proteins before their function as essential elements of the holoenzyme became clear. As such, each sigma factor is known by a variety of names. Designations with *rpo* or Rpo are used across species and refer to the particular stress to which the sigma factor responds. A more contemporary convention is to use a lower case Greek sigma with the molecular weight of the sigma factor as a superscript. In this text, all molecular weights refer to those found in *Salmonella* and *E. coli*.

Most of these proteins, σ^{70}, σ^E, σ^H, σ^S, and σ^F, belong to the same family of sigma factors, potentially all derived from some ancestral form or ancestral regulatory process. The other sigma factor, σ^N, belongs to a different family, although it is the only modern day example found, and may belong to a more ancient regulatory system that has become obsolete with current patterns of growth and reproduction for bacteria like *Salmonella*. While the housekeeping sigma was found to facilitate most gene expression during exponential growth, each of the other sigma factors was found to help the organism address different environmental stresses. Each sigma factor recognizes a different consensus sequence within the promoter region. The relative affinities of multiple sigma factors for the same promoter region determine which recognizes it more often at a specific intracellular concentration.

3.1 σ^N – Nitrogen regulation

σ^N seems to be more evolutionarily distant from the other alternate sigma factors and it may be the remnants of a more ancient regulatory system. In fact, the processes governed by σ^N may not be essential or may be under transcriptional control of another sigma factor (Morett & Segovia, 1993). These processes include nitrogen fixation, dicarboxylic acid transport, and hydrogen oxidation (Morett & Segovia, 1993). Down-regulating expression from RpoN-dependent genes provides increased resistance to killing by host cationic antimicrobial peptides (Barchiesi et al., 2009), indicating that some of these processes may even be detrimental to the organism in certain conditions. In some related species σ^N is related to pathogenicity, but that does not appear to be in *Salmonella* (Studholme, 2002).

The differences between σ^N and the rest of the sigma factors are profound. There is almost no sequence similarity between the *rpoN* gene and genes for other known sigma factors, also suggesting a different origin (Morett & Segovia, 1993). σ^N promoters are unique in that they have conserved consensus sequences centered at -24 and -12 nucleotides from the transcription start site, as opposed to -35 and -10 (Barrios et al., 1999). A highly conserved RpoN-Box is involved in the recognition of the -24 and -12 DNA sequences (Barrios et al., 1999). The distance between the -24 and -12 elements is more stringent than the analogous distance between the -35 and -10 elements for the σ^{70} family of sigma factors, indicating a highly controlled regulation (Barrios et al., 1999). Moreover, the sequences at the -24 and -12 elements have highly conserved GG and GC regions respectively, also suggesting a high level of regulatory control (Barrios et al., 1999).

While the σ^N protein is very different from other alternate sigma factors, the interaction between the sigma factor and template DNA is also distinct. The σ^{70} family of sigma factors do not form stable closed complexes and transcription will start spontaneously (Barrios et al., 1999). Unlike other sigma factors, the σ^N and core enzyme form a stable closed complex. In this way, σ^N binding to the core enzyme actually blocks transcription because the open complex must be activated (Buck & Cannon, 1992). The binding of the RNA polymerase holoenzyme with σ^N as the sigma factor cannot induce DNA melting alone, similar to the RNA polymerase II system in eukaryotes (Buck et al., 2000). σ^N may bind to DNA first rather than binding to the core enzyme first (Buck et al., 2000). This is supported by the fact that σ^N binds to a different location on the core enzyme than σ^{70} and in doing so may be able to assist in DNA melting once activated (Buck et al., 2000).

A Tale of 6 Sigmas: How Changing Partners Allows Salmonella to Thrive in the Best of Times and
Survive the Worst of Times

101

Because it forms a stable closed complex, the RNA polymerase with σ^N as the sigma factor requires enhancer proteins for activation. Each enhancer protein is under the regulation of its own signal transduction pathway, allowing response to various environmental conditions (Buck et al., 2000). All the enhancer proteins have hidden ATPase activity that allows for the DNA melting necessary to initiate transcription (Buck et al., 2000).

3.2 The housekeeping sigma σ^{70}

The other five sigma factors appear to be evolutionarily related, developing from the original or primary sigma factor. RpoD or σ^{70} is the housekeeping sigma factor and is responsible for the transcription of most of the genes in bacterial cells growing exponentially (Ishihama, 1993). When *rpoD* was found in the genome for *E. coli*, it was determined that the gene sequence had a high degree of homology between other *rpoD* genes from closely related species (Scaife et al., 1979). Further genomic analysis determined that *rpoD* is found in a transcript with the 30S ribosomal protein S21 and DNA primase (Burton et al., 1983). This operon was the first discovered operon containing proteins involved in transcription, translation, and replication (Burton et al., 1983). $E\sigma^{70}$ (the holoenzyme containing the core enzyme associated with σ^{70}) does not form a stable closed complex and transcription begins spontaneously (Barrios et al., 1999), requiring no enhancer proteins. Moreover, the σ^{70} concentration found inside a cell undergoing exponential growth is less than the concentration of core enzymes, indicating the level of the sigma factor present may regulate the level of transcription (Burton et al., 1983).

3.3 σ^E – Response to extracytoplasmic stress

When the bacteria face stressors, other sigma factors are involved in the expression of genes necessary to survive the stress, such as σ^E, σ^{24}, or RpoE which results in transcription of genes to combat envelop stress (Kenyon et al., 2005). RpoE is constitutively expressed in the bacteria, held inactive by interaction with various binding proteins. The *rpoE* gene seems to be the most highly conserved of alternate sigma factors across several species, as are the genes under its transcriptional control.

RpoE must be able to respond to a signal coming from outside of the cell, while the protein itself resides inside the bacterium. It appears that a transmembrane protein, RseA, interacts with RseB on the periplasmic side and with σ^E on the cytoplasmic side. An area of the DegS protein on the periplasmic side recognizes unfolded proteins resulting in proteolysis of the periplasmic side of RseA. Cleaved RseA is a target for RseP, which then cleaves the transmembrane portion of RseA, releasing the RseA/ σ^E complex from the membrane and the unstable cytoplasmic portion of RseA is quickly degraded by cytoplasmic proteases (Muller et al., 2009). RseB also interacts on the periplasmic side with both DegS and RseP to control the activity of these proteases in the absence of a stress response (Muller et al., 2009). The strength of the signal is directly proportional to the number of misfolded outer membrane proteins.

While response to envelop stress is typically the signal necessary to release RpoE from RseA, acid stress may also result in the same response. It was found that mutants deficient in RpoE activity showed increased susceptibility to acid and reduced ability to survive inside macrophages. The RseP domain was required for this response to the acid shock, but its proteolytic activity was not dependent on DegS (Muller et al., 2009). It is proposed that the acidic milieu affects the interaction between RseB and RseP, which normally keeps RseP inactive, so that RseP is released to act on RseA, discontinuing negative control over σ^E (Muller et al., 2009). Both DegS and RseP have cytoplasmic and periplasmic domains, and the acid response appears to be independent of the envelope stress response. Again, the response strength is contingent upon the acidity of conditions and the length of exposure.

Once σ^E is released to interact with RNA polymerase, not all σ^E - dependent genes are transcribed equally. Within the approximately 60 promoters examined that required σ^E for transcription, there were few very strong promoters (showing high affinity) but many relatively weak promoters. The strong promoters were conserved across both *E. coli* and *S. enterica*, and were typically involved in maintaining porin homeostasis (Mutalik et al., 2009). Varying strength of promoters allow quick and efficient adaptation to different environments by being able to transcribe different genes in response to various signals (Mutalik et al., 2009). If the stress signal is strong, the cellular concentration of σ^E will increase to transcribe at high rates from weak promoters.

In order to prevent wasted energy and further damage to the cell, the activation of σ^E also results in the down-regulation of *omp* (outer membrane protein) mRNA (Papenfort et al., 2006). The cell also prevents these nascent mRNAs from producing misfolded proteins while avoiding destruction by the exocytoplasmic stress. Two small non-coding RNAs, RybB and MicA, not under the control of RpoE, collectively expedite the destruction of *omp* mRNAs. Under normal conditions, the cellular machinery making OMPs is still not perfect and some misfolded proteins are generated. In this case, the same two sRNAs are involved in the response to fix the problem by inducing the σ^E response, but at a much lower level than would be found in bacteria responding to prolonged stress (Papenfort et al., 2006). As such, the two sRNAs are most likely under the transcriptional control of the housekeeping sigma factor and their increased activity helps to induce σ^E activity.

As far as specific genes governed by σ^E, the parts of the σ^E regulon that are highly conserved across species are involved in making the cell wall and outer membrane of Gram-negative bacteria (Rhodius et al., 2006). The variable portion may be involved in the alternative lifestyles that the studied species utilize. A genome-wide search was done for σ^E-dependent genes in several species including *E. coli* and *Salmonella* Typhimurium, determining that several genes were at the core of the σ^E regulon. Some genes were involved in making lipoproteins, such as *yfiO*, *yeaY*, and *yraP*. Others were involved in outer membrane protein synthesis and modification, like *yeaT*, *skp*, *fkpA*, and *degP*. And still others were involved in cell envelope structure, such as *plsB*, *bacA*, *ahpF*, and *ygiM*. Interestingly, both *rpoE* and *rpoH* were both under regulatory control of σ^E, indicating that σ^E promotes its own transcription and the transcription of other sigma factors (Rhodius et al., 2006). By autoregulation, σ^E can create a multi-fold increase in gene product from its regulon. All of the genes found to be under the control of σ^E are related to making proteins for cellular structure.

3.4 σ^H – Response to heat shock

One of the genes under the transcriptional control of RpoE is another sigma factor, RpoH or σ^{32} (Rhodius et al., 2006). This sigma factor has been found to be involved in the transcription of genes that help *Salmonella* withstand high temperatures, potentially as a result of fever response within the host. Whereas σ^E appears to mediate the response to misfolded outer membrane proteins, σ^H is involved with proteins within the cytoplasm that are misfolded (Bang et al., 2005). Concomitant with increased heat exposure, cell wall and membrane proteins begin to misfold and denature. As the concentration of σ^E increases in response to the misfolded proteins, σ^H also accumulates to respond to a sustained stressor. This is supported by the finding that *rpoH* expression is directly proportional to σ^E activity at temperatures above 42°C (Testerman et al., 2002), a temperature at which protein denaturing begins with the cell.

RpoH governs the transcription of genes such as those encoding proteases that allow for the removal of misfolded proteins within the cytoplasm. For example, an operon composed of *opdA* and *yhiQ* was found to be immediately proceeded by a consensus sequence for the RpoH promoter (Conlin & Miller, 2000). While the function of these two proteins has not been directly studied in the heat shock response, OpdA is metalloprotease oligopeptidase A that would be helpful in degrading misfolded proteins.

Some researchers have also hypothesized that σ^H is related to RNA thermometers, which are other regulatory means for activating and utilizing heat shock genes. RNA thermometers are areas of 5'-untranslated region that fold and complementary pair in such a way as to block the Shine-Dalgarno (SD) sequence of downstream genes (Waldminghaus et al., 2007). When heated to high enough temperatures, these areas unpair to allow the ribosome access to the SD sequence. A previously undescribed RNA thermometer was found within the 5'-UTR of the *agsA* gene in *S. enterica*. This gene is known to be involved in response to heat shock, and has a promoter region that is a consensus sequence for RpoH. Within the *agsA* mRNA appear to be RNA thermometer sequences (Waldminghaus et al., 2007).

In *E. coli*, the *rpoH* mRNA itself contains RNA thermometers. In this species, the cellular level of the RpoH is controlled by complementary base pairing in its mRNA. Unlike other RNA thermometers, the SD sequence is not blocked but the start codon is inaccessible to the ribosome and two halves of the ribosome-binding site pair at low temperatures (Waldminghaus et al., 2007). A similar mechanism is likely at play in *Salmonella*.

While responding to heat shock is vitally important for survival of the bacteria, the most important function of σ^H is to mediate σ^E regulation of σ^S through *hfq* gene expression. In *E. coli*, the promoter sequence found upstream of the *hfq* gene was found to be σ^H –dependent. The same promoter was found in *S.* Typhimurium (Bang et al., 2005). In conditions with scarce nutrients, σ^E appears to upregulate σ^S through the increase of σ^H (Bang et al., 2005).

The product of the *hfq* gene, HF-I, is important for translation of RpoS. This small protein is heat stable and binds to RNA to facilitate translation (Brown & Elliot, 1996) by associating with the ribosome (Brown & Elliot, 1997). Several possible mechanisms for the manner in which the protein encoded by *hfq* regulated σ^S translation have been suggested, including preventing the interaction of some sort of antisense mRNA or by being directly involved in the transcription of *rpoS* (Cunning & Elliot, 1999). Most evidence supports the assertion that the function of HF-I is as an RNA chaperone, after it was demonstrated to bring the mRNA and ribosome in correct association for translation (Sittka et al., 2007).

3.5 σ^S – Stationary phase growth, response to stress, and response to starvation

The role of this sigma factor, also called σ^{38}, is slightly more difficult to define than that of RpoE or RpoH. However, it is clear that the function of RpoS is essential. The conserved sequence of *rpoS* across multiple species and within the same species found in different geographical areas speaks to its importance. When *rpoS* genes are characterized in clinical isolates, the mutations found are not clonal but rather novel, implying that there is some selection against mutants. Even when strains demonstrated different abilities to survive certain stresses like exposure to hydrogen peroxide, it did not appear to be related to different *rpoS* genes (Robbe-Saule et al., 2007).

The number and types of genes that seem to be under transcriptional control of σ^S have a variety of functions and respond to a wide variety of lifestyle requirements and threats to survival. The only known constant about the genes transcriptionally governed by RpoS is their dependence on growth phase (Ibanez-Ruiz et al., 2000). Previously, work has determined that during logarithmic growth, any activity from σ^S promoters is repressed by cyclic-AMP receptor activity (Fang et al., 1996). Stationary phase growth is characterized by a lack of cellular multiplication and decreasing cell density. The transition from exponential growth to stationary phase growth is the result of the concentration of a regulatory protein (Hirsch & Elliot, 2005). The concentration of Fis (factor for inversion stimulation), a DNA binding protein, is high during exponential growth and low in stationary phase. Fis binds to a region of DNA upstream of the promoter for *rpoS* and with decreasing concentration, allows the switch to stationary phase (Hirsch & Elliot, 2005).

A genome-wide search has been done for genes under the transcriptional control of RpoS. The project found that, like RpoE, the σ^S regulon includes promoters of various strengths. Despite the assumed similarities between the *E. coli* and *S.* Typhimurium genome, there were several genes within the *Salmonella* genome that were not homologous with any genes of *E. coli*. Several genes of unknown function were found under the control of σ^S, as was *ogt*, which encodes the enzyme O^6-methylguanine DNA methyltransferase (Ibanez-Ruiz et al., 2000). This enzyme is responsible for repairing DNA damaged by alkylation (Fang et al., 1992).

σ^S also seems to play a role in a wide variety of other functions that ensure the survival of the bacteria, such as protection from acid shock and nutrient depletion. Decreased pH unfolds the secondary structure stem and loops of the *rpoS* mRNA, allowing availability for translation (Audia & Foster, 2003). Constitutive degradation of the sigma factor coupled with no more being made results in the system reset after the acid threat has passed (Audia & Foster, 2003). RpoS also seems to be involved in survival of the bacteria in starvation conditions. σ^S has been found to act as both a positive regulator for *stiA* and *stiC* and a negative regulator for *stiB*. These three genes are part of the multiple-nutrient starvation-induced loci. σ^S was required for phosphate, carbon, and nitrogen starvation survival through induction of *stiA* and *stiC*. σ^S also acted as a negative regulator of *stiB* during phosphate and carbon starvation induced stationary growth (O'Neal et al., 1994).

3.6 σ^F – Flagellar formation and chemotaxis

Flagellar assembly was originally assumed to be under the control of σ^{70}, because it seemed essential to survival. However, examining promoters of known flagellar genes found no consensus sequences for σ^{70} (Helmann & Chamberlin, 1987). Instead, researchers found promoter sequences in *Salmonella* known to be used by alternative sigma factors in closely related species (Helmann & Chamberlin, 1987). σ^F, more commonly called FliA, or σ^{28}, has the most specific function of all the alternate sigma factors. FliA is involved in the transcription of genes related to the formation of flagella, specifically the formation of the

A Tale of 6 Sigmas: How Changing Partners Allows Salmonella to Thrive in the Best of Times and
Survive the Worst of Times

105

flagellar filament (Ohnishi et al., 1990). Operons of flagellar assembly are proceeded by one of three classes of promoters, class 1, 2, or 3 (Bonifield & Hughes, 2003, Karlinsey et al., 2000, Karlinsey et al., 2006) which allow for a temporal regulation of gene expression. From these operons, more than 50 genes are transcribed to allow complete flagellar assembly and function (Kutsukake et al., 1994).

There is only one class 1 operon which encodes the *flhD* and *flhC* genes (Karlinsey et al., 2000). Class 1 is the master operon, with FlhD and FlhC acting as a global regulator of flagellar assembly (Karlinsey et al., 2006). FlhD and FlhC form a heterotetrameric complex that is a positive transcriptional activator of class 2 promoters through σ^{70}, by interacting with the α subunit of the core enzyme (Bonifield & Hughes, 2003, Liu et al., 1995, Liu & Matsumura, 1994). Class 2 operons include genes for the assembly of the hook and basal body complex (HBB), σ^F, and FlgM (Bonifield & Hughes, 2003). The basal body, containing the motor, penetrates the cell membrane and includes the hook element on the extracellular side of the cell (Brown & Hughes, 1995). The filament protrudes from the hook into the extracellular matrix and turns to provide motility.

The third class of flagellar operons requires σ^{28} or FliA for transcription (Bonifield & Hughes, 2003). Proteins generated from these operons are for the flagellar filament, the generation of motor force, and chemotaxis (Karlinsey et al., 2006). FlgM, which is also transcribed from class 2 operons along with FliA, acts as an anti-sigma factor, keeping FliA inactive until the completion of the HBB. The C-terminal portion of FliA has a binding site for FlgM (Kutsukake et al., 1994). FlgM prevents RNA polymerase core enzyme from interacting with FliA to transcribe class 3 flagellar operons (Chadsey et al., 1998). The FlgM protein is able to assess the completion of the HBB because the protein itself is an exported substrate (Hughes et al., 1993). Decreasing concentrations of FlgM release FliA to interact with the RNA polymerase core enzyme and transcribe class 3 operons (Hughes et al., 1993). The relative concentration of FliA to FlgM determines the number of flagella that a single cell will have (Kutsukake & Iino, 1994). Additionally, the FlhD/FlhC complex may assist FliA in association with the RNA polymerase (Kutsukake & Iino, 1994). FlhD is involved in assessing nutrient state (Chilcott & Hughes, 2000), which may be requisite for bacterial motility.

The intracellular concentration of FliA and FlgM is governed by other regulatory mechanisms as well. The genes from both of these proteins can be transcribed from either class 2 or class 3 promoters (Wozniak et al., 2010). In this way, FliA can positively and negatively regulate its own intracellular concentration dependent upon the concentration of FlgM within the cell (Ikebe et al., 1999). Mutants lacking FlgM overproduce flagella via overexpression from class 3 operons (Yokoseki et al., 1996).

4. Changing partners

The presence of alternate sigma factors has been well studied, but how do the alternate sigma factors displace the housekeeping sigma or each other to govern gene transcription? Most of the answer points to concentration dependence; that is, the concentration of a particular sigma factor changes in response to different environmental conditions. For example, RpoE, as discussed above, is expressed constitutively but held inactive by various other proteins until an extracellular signal is received. This signal activates a series of proteolytic activities that gradually increases the intracellular concentration of RpoE. Once RpoE is released, it is free to interact with the core enzyme. RpoE is positively autoregulated and as genes are transcribed from RpoE-response promoters, the intracellular concentration increases exponentially so that the intracellular concentration of RpoE can outcompete other sigma factors for binding access to the core enzyme. RpoE, in turn, allows for transcription of *rpoH*, which summarily mediates *rpoS* expression, increasing the intracellular level of all three alternative sigma factors. Fine tuning of these concentrations allows for precise control of gene expression. If a finite amount of RNA polymerase is available, increasing the presence of one sigma factor can repress expression of genes requiring a different sigma factor (Farewell et al., 1998).

Growth phase also appears to play a role in the intracellular concentration of certain sigma factors. During exponential growth, intracellular concentrations of σ^{70} remain relatively constant and σ^S is basically absent (Jishage & Ishihama, 1995). During stationary phase growth, the intracellular concentration of σ^S increases to nearly 30% of σ^{70} concentration (Jishage & Ishihama, 1995). Moreover, the concentration of the core enzyme decreases during stationary phase growth (Jishage & Ishihama, 1995), meaning that a 30% increase in concentration is more than a 30% increase in competitive advantage. RpoS activity is repressed by the products of *uspA* and *uspB*, which are both under the transcriptional control of σ^{70} (Farewell et al., 1998). During exponential growth, σ^S is highly unstable (Jishage & Ishihama, 1995). In stationary phase growth, σ^S is released and free to interact with RNA polymerase core enzyme. Researchers have hypothesized that there may be a σ^{70} anti-sigma factor under transcriptional control of σ^S or that a change in the cytoplasm may favor σ^S – mediated transcription (Farewell et al., 1998). Most genes expressed during exponential growth are not expressed during stationary phase growth, so σ^{70} proteins need to be rendered inactive (Jishage & Ishihama, 1995). Interestingly, the intracellular levels of σ^S reach those of σ^{70} during osmotic shock (Jishage & Ishihama, 1995), indicating that the change in concentration of a sigma factor can be a gradual or dramatic.

Environmental conditions can also play a role in the stability of the proteins, which can affect transcriptional efficiency. For example, RpoH, the heat shock sigma factor, is highly unstable at low temperatures; but, above 42°C intracellular concentrations will transiently increase (Jishage & Ishihama, 1995). Higher temperatures may provide increased efficiency of σ^H – mediated transcription or they may stabilize the protein itself so that it is able to interact with the core enzyme (Jishage & Ishihama, 1995).

5. *Salmonella* as pathogenic bacteria

In determining how alternate sigma factors are able to promote survival and spread of *Salmonella*, it is important to understand how *Salmonella* lives. *Salmonella* typically enters its host through the oral route. If sufficient numbers are ingested, some organisms will survive the low pH conditions of the stomach to reach the small intestine (Dougan et al., 2011). Sometimes the bacterial infection is halted here. For a systemic infection to occur, the bacteria must invade the gut epithelium (Hansen-Wester & Hensen, 2001). *Salmonella* preferentially invade epithelial cells in the distal ileum of the small intestine by adhering to and then injecting effector proteins into the host cell that facilitates bacterial entrance into membrane bound vesicles (Bueno et al., 2010). The small intestine provides an environment of near-neutral pH and high osmolarity, condusive to invasion not found in the large intestine (Lawhon et al., 2002).

Within the small intestines, *Salmonella* specifically invades Peyer's patches through M cells. Peyer's patches are specialized lymphoid tissues that are designed to sample intestinal antigens and lead to immune responses (Slauch et al., 1997). *Salmonella* exclusively enter M cells found within the follicle-associated epithelium of Peyer's patches (Jones & Falkow, 1994). M cells are epithelial cells responsible for the uptake of luminal antigens (Slauch et al., 1997) and can engulf large particles, making them ideal for target by *Salmonella* (Jones & Falkow, 1994). When one bacterium makes entry into the host epithelial cell, it recruits other pathogens to its location (Francis et al., 1992).

6. Islands of pathogenicity

An estimated 5-10% of genes within the *Salmonella* genome can be considered virulence genes (Slauch et al., 1997). These genes have been found arranged in clusters within the *Salmonella* chromosome, the so called *Salmonella* Pathogenicity Islands (SPIs). It has been theorized that these gene clusters were acquired by horizontal transfer based on their higher G-C content compared with other parts of the *Salmonella* chromosome (Slauch et al., 1997) and because similar regions are not found in closely related commensal species such as *E. coli* (Galán, 1996). There are at least five known SPIs, but SPI-1 and SPI-2 seem to the most important in the initial phases of infection. Both SPI-1 and SPI-2 encode type III secretion systems (TTSS) (Shea et al., 1996). Additionally, genes within the SPIs encode effector proteins and regulatory proteins (Hansen-Wester & Hensen, 2001). These secretion systems allow the insertion of effector proteins into the extracellular environment and inside the host cell.

SPI-1 appears to contain genes involved in bacterial uptake by the host cell, while SPI-2 genes are involved in survival inside cells (Lara-Tejero & Galán, 2009). However, there is some evidence indicating that SPI-1 may also be important for bacterial life inside the vacuole and for their survival and replication intracellularly (Steele-Mortimer et al., 2002). Secreted proteins from genes transcribed from SPI-1 leads to actin cytoskeleton rearrangement of the host cell that facilitates bacterial entrance into membrane bound vesicles (Chen et al., 1996). Once inside the cell, a variety of functions can be hijacked to serve the bacteria's purpose, including cytoskeleton arrangement, vesicular trafficking, cell cycle progression, and programmed cell death (Lara-Tejero & Galán, 2009). These effector proteins activate GTP-binding proteins such as Cdc42, Rac-1, and Rho, which coordinate intracellular activities in the host cell (Chen et al., 1996). Effector proteins also down-regulate actin rearrangement (Fu & Galán, 1999) to reverse the actin rearrangement.

Transcription of all SPI-1 operons is activated by a regulatory loop beginning with HilA (Matsui et al., 2008). Through other regulator proteins like HilC, HilD, and InvF, expression of invasion genes is modulated with HilA as the central player (Lucas et al., 2000). Interestingly, the rising concentration of acetate in the distal intestine activates the expression of HilA, bypassing normal positive regulators (Lawhon et al., 2002).

While SPI-1 may play a role in the procession of the infection past the initial invasion of epithelial cells, SPI-2 is vital for the migration of the bacteria to other parts of the host (Löber et al., 2006). SPI-2 was the second pathogenicity island discovered and is required for virulence after the bacteria has entered into the epithelial cells (Shea et al., 1996). This claim is further supported by evidence that mutants without SPI-2 genes could enter Peyer's patches but were unable to spread to mesenteric lymph nodes (Cirillo et al., 1998). Not all members of the SPI-2 pathogenicity island are equally vital for the ability of the pathogen to establish systemic infection. Mutants with various genes knocked out show a varying level of attenuation (Cirillo et al., 1998, Hensel et al., 1998). However, the genes within the SPI-2 are responsible for avoiding destruction by lysosomes within dendritic cells and macrophages (Tobar et al., 2006). Expression of SPI-2 genes seems to be induced by the slightly acidic conditions inside the initial vacuole formed when the bacteria are initially internalized by the host cell (Löber et al., 2006).

6.1 Regulation of *Salmonella pathogenicity islands* by sigma factors

Regulatory control of SPIs can be exerted by sigma factors without sigma factors being directly involved in the transcription of these genes. SPI-1 genes are typically transcribed using σ^{70}. σ^H mediates SPI-1 expression by regulating activators of SPI-1. Systems mediated by RpoH negatively regulate HilD post-translationally and HilA transcriptionally (Matsui et al., 2008). HilD is responsible for activating HilA transcription, and HilA in turn activates all the genes within SPI-1. σ^H directs the production of Lon protease which specifically degrades HilD (Matsui et al., 2008). By modulating the activation of σ^H, the bacterial cell can control SPI-1 expression, restricting expression to specific regions within the host cell (Matsui et al., 2008). The cell can repress invasion genes long enough to replicate, escape, and invade a new macrophage before cell death (Matsui et al., 2008).

Promoters for SPI-2 genes all have consensus sequences for σ^{70} (Osborne & Coombes, 2009). However, upstream of some SPI-2 genes seem to be consensus sequences for σ^E recognition (Osborne & Coombes, 2009). It is postulated that these σ^E binding sites may serve a couple of different purposes. The σ^E – recognized promoters may allow the bacteria to express TTSS in response to host factors that compromise bacterial cellular integrity (Osborne & Coombes, 2009). Alternatively, σ^E may fine-tune the expression of SPI-2 genes through σ^{70} (Osborne & Coombes, 2009) by preferentially overexpressing certain genes while all others are expressed at basal levels by σ^{70}.

Stationary phase *Salmonella* are unable to cause actin rearrangement in the host epithelial cell that is necessary for entry (Francis et al., 1992). Invasion factors are either not functional or not expressed in stationary phase bacteria (Francis et al., 1992). As growth phase has been demonstrated to change intracellular concentrations of different sigma factors and virulence genes do not appear to be under the transcriptional control of σ^S, it stands to reason that these bacteria would not be able to invade; invasion genes would be inactive since the activity of the necessary sigma factor is repressed.

6.2 Other genetic sources of virulence

Virulence genes may be found outside of *Salmonella* pathogenicity islands. These genes are similarly essential to survival and also are responsive to changes in sigma factor availability. While the genetic location of the Spv regulon varies among *Salmonella* species from chromosomal to plasmid-encoded, all species carry the regulon and it functions to increase intracellular growth in host cells once the bacteria have spread outside of the small intestine (Guiney et al., 1995). σ^S mutants are unable to efficiently express the Spv regulon. Expression of one of the members of the Spv regulon, *spvB*, decreased by 86% when σ^S was knocked out (Fang et al., 1992). The lethal dosage in mice for a strain without a functional *rpoS* gene was 1000 fold greater than the wild type (Fang et al., 1992).

The dependence of Spv regulon expression on growth phase also indicates a dependence on
σ^S for transcription. However, it seems to be nutrient availability, not cell density, that is
most important in mediating Spv regulon expression (Guiney et al., 1995). σ^S associated with
RNA polymerase results in expression of genes that are essential to help the bacteria survive
nutrient depleted conditions, such as those found in deeper tissues beyond the small
intestine (Guiney et al., 1995).

σ^S increases expression of *spv* virulence genes by interacting with SpvR, a repressor protein
for the virulence plasmid (Kowarz et al., 1994). Competition for RNA polymerase between
σ^S and σ^{70} led to less efficient transcription of *spvR* from its promoter as σ^S has a greater
affinity for RNA polymerase than σ^{70} but a lower affinity for the promoter for *spvR* (Kowarz
et al., 1994). σ^S affinity for RNA polymerase is enhanced by its interaction with the Crl
protein, giving it the ability to displace σ^{70} as the preferred promoter (Robbe-Saule et al.,
2007). The presence of SpvR regulates its own transcription (Kowarz et al., 1994) so the lack
of efficient transcription leads to decreasing cellular levels and derepression of *spv* plasmid
virulence genes. σ^S ensures that enough SpvR is present to activate transcription from the
spvA promoter, the first gene in the regulon (Guiney et al., 1995).

7. Sigma factors and surviving the best of times and the worst of times

While *Salmonella* Pathogenicity Islands allow the bacteria to invade host cells, the pathogen
must then survive the hostile environment found inside. While differential gene expression
from various sigma factors ensures the appropriate expression of SPIs to gain access to the
intracellular milieu of the host, the use of alternate sigma factors also permits survival.

7.1 A sigma factor cascade for survival in phagocytic cells

Ferric Fang describes a cascade of transcriptional and translational events that involve sigma
factors associating with the core enzyme to transcribe genes for each other and those
necessary to respond to a variety of assaults in the intracellular environment (Fang, 2005).
The first step in the cascade is activation of σ^E, which is constitutively expressed through σ^{70}
promoters, but held inactive by a pair of negative regulators, RseA and RseB (De Las Peñas
et al., 1997). RseA interacts with σ^E in such a way as to block the binding site for RNA
polymerase (Muller et al., 2009). When an extracytoplasmic stress is perceived, σ^E is released
by RseA and freed to bind to RNA polymerase. Interaction of σ^E with the core enzyme
allows for transcription from other promoters. These promoters include those before the σ^E
regulon of genes but also before the *rpoH* gene, which encodes the alternative sigma factor,
σ^H. σ^H provides specificity for RNA polymerase to transcribe genes in the σ^H regulon, which
respond to cytoplasmic stress. Additionally, σ^H allows transcription of *hfq*. The Hfq protein
interacts with the *rpoS* mRNA to facilitate its translation. The σ^S then allows transcription of
genes under its transcriptional control, which allow for a starvation response (Fang, 2005).
This overall cascade allows for coordinated response by the pathogen. To ensure that sigma
factors help transcribe genes needed to respond to stress only as long as it exists, there must
be some mechanism of turnover (Fang, 2005). In this way, the use of an interconnected web
of sigma factors allows *Salmonella* to gain access to various cell types and then survive to be
able to spread to other areas of the host.

This cascade's vital importance to survival, in particular within macrophages, is illustrated
by the increased levels of σ^S inside the macrophage following infection. Some aspect of

being inside a macrophage results in increased transcription of the *rpoS* gene. While levels of the housekeeping sigma σ^{70} decreased, levels of σ^{S} increased about 10-fold a few hours after infection (Khan et al., 1998). Conditions inside the macrophage induce the stress response and restrict nutrient availability, which induces the sigma cascade of gene expression to help the bacteria survive, although not necessarily to increase/induce virulence.

7.2 Sigma factors coordinate gene expression

Rarely is gene expression controlled in a strictly linear manner. That is, multiple sigma factors may work together to fine tune an expression of a group of genes to provide the bacteria with high probability of survival. The cascade of sigma factors used to allow survival inside phagocytic cells (described above) is just one example. There are many other instances of sigma factors working simultaneously.

One way to determine if one sigma factor plays a role in the efficient transcription by the other is to knock out one of them and see how the function of gene products mediated by the other are affected. In this way, investigators determined a relation between RpoE and FliA. Mutants without *rpoE* showed defective or limited mobility (Du et al., 2011). In these mutants, expression from class 1 flagellar promoters remained unaffected while some class 2 and most class 3 promoters showed decreased activity as compared to wild type (Du et al., 2011). It was concluded that RpoE may promote expression from class 3 promoters by mediating expression of FliA during osmotic stress, such as the hyperosmotic conditions found in the small intestine (Du et al., 2011).

RpoH and RpoN also appear to be related based on their ability to control the same genes as well as their dependence on one another. Expression of some heat shock operons appear to be under the control of RpoN in certain conditions, as expression from σ^{H} operons is down-regulated in mutants with an *rpoN* knockout (Studholme, 2002). In this way, RpoN may be responsible for fine tuning some gene expression during heat shock response. The expression of topoisomerases also appears to be governed by both σ^{N} and σ^{H} (Studholme, 2002), which may also indicate an interdependence of the activities of the two sigma factors.

Insufficient expression of one sigma factor can be compensated for by over-expression of other sigma factors. For example, researchers expected that because RpoS was vital to survival within macrophages, this sigma factor would be important for expressing virulence genes inside these phagocytic cells. However, within macrophages while RpoS only moderately increased following infection, RpoH and RpoE showed dramatic increases in intracellular concentration (Eriksson et al., 2003). While RpoS is typically associated with virulence inside phagocytic cells, it may be possible for other sigma factors to express other genes in response to a different environmental stimulus while still ultimately resulting in virulence. Research has also demonstrated that RpoN can compensate for insufficient RpoS in the formation of certain lipopolysaccharides (Bittner et al., 2004).

7.3 Survival outside of a host

While *Salmonella* is an important enteric pathogen to study because it infects many hosts and can be transmitted from species to species, it also is able to survive for long periods of time outside a host. Because of this characteristic, it has been an important target of sanitation processes to eliminate possible sources of transmission.

One mean of *Salmonella* transmission to human hosts is through food products, such as poultry. The same mechanisms of alternate sigma factor used to survive acid challenges in the mammalian gut are also utilized in surviving the fowl gastrointestinal tract and can lead to transmission of the pathogen (Dunkley et al., 2008).

Other studies specific to food handling procedures and alternate sigma factors have determined that RpoS, for example, is essential to *Salmonella*'s ability to withstand normal sanitation procedures common in the food service industry and that early induction of RpoS can cause the cells to enter stationary growth phase prematurely, negating the protective nature of stationary growth on the pathogen's ability to survive (Komitopoulou et al., 2004). Other studies have demonstrated that certain food handling processes, such as washing in various antimicrobial agents, can induce RpoS to protect the bacteria from destruction (Dodd & Aldsworth, 2002). Significant drops in temperature have also been found to activate transcription from σ^S dependent promoters rather than from the σ^{70} promoters from which genes are normally transcribed (Rajkumari & Gowrishankar, 2001).

Multiple alternate sigma factors contribute to survival through food processing. For example, σ^S and σ^E were both found to be important in surviving refrigeration and changes in osmotic pressure. Depending on the nature of the stress, either σ^S or σ^E may be more important and their relative concentrations dictate the response (McMeechan et al., 2007).

8. Transcriptional and translational regulation of sigma factors

Because sigma factors are capable of effecting dramatic changes in cellular protein composition and energy use, their actions must be closely guarded to ensure that the pathogen is responding to the stress without exhausting cellular resources.

8.1 Regulation of sigma factors

Some alternate sigma factors are constitutively expressed but held inactive until they are needed by regulatory proteins that change conformation or leave the cell in response to a particular signal. RpoE is held inactive until an extracellular signal of extracytoplasmic stress is received and FliA is held inactive by FlgM until the FlgM is exported out of the cell by the completed hook and basal body structure. Some regulation of sigma factors is accomplished by the optimal conditions under which they can influence gene expression. *rpoH* cannot be translated below a certain temperature because at lower temperatures the mRNA folds back on itself blocking the start codon. And RpoS shows increased efficiency during stationary phase growth and is almost nonexistent during exponential growth.

Because much of the efficiency of sigma factors to influence transcription is itself influenced by their relative concentrations within the cells, many mechanisms to regulate them change the available concentration of these proteins. Different proteases target specific sigma factors and depending on the relative concentration of these proteases, the relative availability of the sigma factors can be adjusted.

RpoS is needed to transcribe the most genes and is therefore the most highly regulated. Several novel pathways of regulation have been discovered. DksA is required for efficient translation of *rpoS* but not as an RNA chaperone (Webb et al., 1999). Another protein, RstA, decreases the expression of RpoS controlled genes and appears to decrease cellular levels of

RpoS independently of proteolytic activity (Cabeza et al., 2007). Translation of the *rpoS* mRNA is elevated in the presence of appropriate carbon sources, indicating a growth rate dependent control of sigma factor availability (Cunning & Elliot, 1999). In response to increased glucose levels, StpA prevents overactivation of σ^S indirectly enhancing its turnover (Lucchini et al., 2009). Some small mRNAs such as DsrA and RprA, are highly conserved as are their antisense elements within the *rpoS* mRNA, but they only have small effects on RpoS availability (Jones et al., 2006). DsrA interaction with *rpoS* mRNA disrupts the stem and loop base pairing of *rpoS* mRNA to allow high levels of translation (Majdalani et al., 2001). The same study discovered another small RNA, RprA, that interacts in a similar way to positively regulate RpoS translation (Majdalani et al., 2001).

8.2 Sigma factors and other regulatory mechanisms

Differential gene expression through alternate sigma factors is far from the only regulatory mechanism found in *Salmonella*. When these other regulatory systems respond to environmental stimuli, alternate sigma factors influence gene expression related to these systems as well. Two important regulators that intersect differential gene expression with sigma factors are the PhoP/PhoQ regulatory system and the Fis global regulator.

The PhoP/PhoQ regulatory system influences the expression of many genes and is functionally a sensor of extracellular magnesium concentration. It has been hypothesized to have evolved differently in closely related species like *E. coli* and *Salmonella* as a result of different lifestyles (Monsieurs et al., 2005). The relation between the PhoP/PhoQ regulatory system and σ^S appears to be essential. Even in cells with functional copies of *rpoS*, mutants lacking PhoP cannot form functional phagosomes within phagocytic cells (Alpuche-Aranda et al., 1994). Mutants with a double knockout of the RpoS and PhoP/PhoQ show decreased virulence and decreased invasion of host cells (Lee et al., 2007). It has even been suggested that because of their inability to cause lasting infections, these double knockouts should be used to make a *Salmonella* vaccine (Lee et al., 2007).

PhoP controls the level of available RpoS by controlling proteins, which enable its degradation by ClpXP. PhoP acts as a transcriptional activator for *iraP*, which encodes a protein that interacts with RssB. RssB facilitates ClpXP degradation of σ^S (Tu et al., 2006). By blocking RssB activity, the level of σ^S accumulates during PhoP/PhoQ activation, which includes low levels of magnesium as found inside macrophages. This is very different than the type of regulation seen in the commensal *E. coli* (Tu et al., 2006), indicating that while there is some similarity in the genes expressed between the two, the regulation of the alternative sigma pathways is not the same. Interestingly, RpoE seems to be involved in the regulation of PhoP/PhoQ activity through Hfq, the same RNA chaperone through which it mediates RpoS expression (Coornaert et al., 2010).

Fis (factor for inversion stimulation) is a global transcription regulator and facilitates site-specific DNA recombination (Mallik et al., 2004). The intracellular concentrations of Fis are high during exponential growth and low in late exponential and stationary phase growth (Walker et al., 2004). The *fis* promoter itself is of some interest as to how these concentration differences are maintained. The σ^{70} dependent and growth-phase dependent regulation from this promoter is achieved through a weak -35 sequence, a second RNA polymerase binding site, and the relative concentration of nucleotides within the cell (Walker et al., 1999). The *fis* promoter is somewhat unique among σ^{70} – dependent promoters in that transcription begins with a cysteine (Walker et al., 2004). This residue is normally a poor initiator of transcription and as a result the RNA polymerase holoenzyme binds very weakly with the *fis* promoter (Walker et al., 2004). When cellular concentrations of cysteine are low, there is very little transcription from the promoter but as CTPs increase in the cell, so does gene expression from the *fis* promoter (Walker et al., 2004).

As expected from the pattern of Fis concentration in the cell, there is a negative relationship between the intracellular level of RpoS and Fis during stationary phase growth (Cróinín & Dorman, 2007). Fis in fact is able to mediate expression from σ^S – dependent genes by binding to a Fis-specific site upstream of σ^S promoter regions and blocking RpoS activity during exponential growth (Hirsch & Elliot, 2005).

Fis, as its name suggests, is also essential for the ability of *Salmonella* to switch flagellar types. There are two types of flagellar filaments, FljB and FliC, which are both transcribed from class 3 promoters. Flagellar switching is achieved by inversion of a promoter region. When expression occurs from this promoter, a type B filament is produced and a repressor of type C is created. When the inversion occurs, the repressor of type C is not produced and type C filaments are made (Aldridge et al., 2006). Hin (for H invertase) and Fis are both required for proper inversion (Bruist et al., 1987). Hin seems to mediate the inversion while Fis ensures the appropriate alignment of the inverted DNA (Bruist et al., 1987).

In having two different types of filaments available for use, *Salmonella* is able to evade the host immune system. FliC is a well-studied target of the immune system (Cummings et al., 2005). As bacteria migrate through the small intestine and into the rest of the host, FliC expression is suppressed or switched for FljB expression to avoid detection by T cells (Cummings et al., 2005). Once past the initial site of infection, T cells are no longer able to recognize the pathogen (Cummings et al., 2005).

Finally, the relatedness of alternate sigma factors and pathogenicity can ensure that certain genes are not expressed at the wrong time. The gene *hilA* which is responsible for the regulation of SPI-1 genes is found in the same operon as FliA, the alternate sigma factor for flagellar filament assembly (Lucas et al., 2000). This proximity within the genome allows for the simultaneous control of both mobility and invasion properties, and ensures the likely co-inheritance of the regulatory elements.

9. Conclusion

Differential gene expression through the use of alternate sigma factors is one of numerous regulatory methods available to *Salmonella* to avoid destruction by its host's immune system or sanitation processes and to thrive in a variety of environments. Control through sigma factors intersects control exerted by other regulatory pathways to ensure a highly controllable pattern of gene expression. The full capacity of *Salmonella* to change rapidly and accurately to respond to environmental conditions is still not well understood. Genes that are under the most types of regulatory control are typically the most important in virulence (McDermott et al., 2011) and it is clear that not only are sigma factors highly controlled themselves at the level of transcription and translation, but they are interconnected in a complex web.

From a medical standpoint, rendering *Salmonella* essentially commensal by knocking out various genes for sigma factors may be an area of interest in creating vaccines. *Salmonella* mutants with one or more nonfunctional copies of genes for alternate sigma factors show significantly attenuated growth across hosts and especially in macrophages, which seems to be the most essential characteristic of *Salmonella's* ability to evade the host immune system. Understanding how sigma factors protect the integrity of the bacteria and testing the limits of this protection may provide insight into the development of new sanitation processes that eliminate more of the bacteria and prevent spread.

10. Acknowledgment

This work was supported, in part, by an IRIG-CEG grant from the office of the Executive Vice President for Research at the University of Louisville.

11. References

Aiba, A. & Mizobuchi, K. (1989). Nucleotide Sequence Analysis of Genes *purH* and *purD* Involved in the *de Novo* Purine Nucleotide Biosynthesis in *Escherichia coli. Journal of Biological Chemistry*, Vol.264, No.35, (December 1989), pp. 21239-21246, ISSN 0021-9258.

Aldridge, P.D.; Wu, C.; Gnerer, J.; Karlinsey, J.E.; Hughes, K.T.; & Sachs, M.S. (2006). Regulatory protein that inhibits both synthesis and use of the target protein controls flagellar phase variation in *Salmonella enterica. Proceedings of the National Academy of Sciences*, Vol.103, No.30, (July 2006), pp. 11340-11345, ISSN 0027-8424

Alpuche-Aranda, C.M.; Racoosin, E.L.; Swanson, J.A.; & Miller, S.I. (1994). *Salmonella* Stimulate Macrophage Macropinocytosis and Persist within Spacious Phagosomes. *Journal of Experimental Medicine.* Vol.179, No.2, (February 1994), pp. 601-608, ISSN 0022-1007

Audia, J.P. & Foster, J.W. (2003) Acid Shock Accumulation of Sigma S in *Salmonella enterica* Involves Increased Translation, Not Regulated Degradation. *Journal of Molecular Microbiology and Biotechnology.* Vol.5, No.1, (March 2003), pp. 17-28, ISSN 1464-1801

Bang, I.S.; Frye, J.G.; McClelland, M.; Velayudhan, J.; & Fang, F.C. (2005) Alternative sigma factor interactions in *Salmonella*: σ^E and σ^H promote antioxidant defenses by enhancing σ^S levels. *Molecular Microbiology.* Vol.56, No.3, (March 2005), pp. 811-823, ISSN 1365-2958

Barchiesi, J.; Espariz, M.; Checa, S.K.; & Soncini, F.C. (2009) Downregulation of RpoN-controlled genes protects *Salmonella* cells from killing by the cationic antimicrobial peptide polymyxin B. *FEMS Microbiol Letters.* Vol.291, No.1, (February 2009) pp. 73-79, ISSN 0378-1097

Barrios, H.; Valderrama, B.; & Morett, E. (1999) Compilation and analysis of σ^{54} – dependent promoter sequences. *Nucleic Acid Research.* Vol.27, No.22, (November 1999), pp. 4305-4313, ISSN 0305-1048

Baumler, A.J.; Kusters, J.G.; Stojiljkovic, I.; & Heffron, F. (1994), *Salmonella* Typhimurium Loci Involved in Survival within Macrophages. *Infection and Immunology*, Vol.62, No.5, (May 1994), pp. 1623-1630, ISSN 0019-9567

Bittner, M.: Saldías, S.: Altamirano, F.; Valvano, M.A., & Cantreras, I. (2004). RpoS and RpoN are involved in the growth-dependent regulation of *rfaH* transcription and O antigen expression in *Salmonella enterica* serovar typhi. *Microbial Pathogenesis*, Vol.36, No.1, (January 2004), pp. 19-24, ISSN 0882-4010

A Tale of 6 Sigmas: How Changing Partners Allows Salmonella to Thrive in the Best of Times and
Survive the Worst of Times

115

Bonifield, H.R. & Hughes, K.T. (2003). Flagellar Phase Variation in *Salmonella enterica* Is Mediated by a Posttranscriptional Control Mechanism. *Journal of Bacteriology,* Vol.185, No.12, (June 2003), pp. 3567-3574, ISSN 0021-9193.

Brown, K.L. & Hughes, K.T. (1995) The role of anti-sigma factors in gene regulation. *Molecular Microbiology,* Vol.16, No.3, (May 1995), pp. 397-404, ISSN 0950-382X

Brown, L. & Elliot, T. (1996). Efficient Translation of the RpoS Sigma Factor in *Salmonella* Typhimurium Requires Host Factor I, an RNA-Binding Protein Encoded by the *hfq* Gene. *Journal of Bacteriology* Vol.178, No.13, (July 1996), pp. 3763-3770, ISSN 0021-9193

Brown, L. & Elliot, T. (1997). Mutations that Increase Expression of the *rpoS* Gene and Decrease Its Dependence on *hfq* Function in *Salmonella* Typhimurium. *Journal of Bacteriology,* Vol.179, No.3, (February 1997), pp. 656-662, ISSN 0021-9193

Bruist, M.F.; Glasgow, A.C.; & Johnson, R.C. (1987). Fis binding to the recombinational enhancer of the Hin DNA inversion system. *Genes and Development,* Vol.1, No.8, (October 1987), pp. 762-772, ISSN 0890-9369

Buck, M. & Cannon, W. (1992). Specific binding of the transcription factor sigma-54 to promoter DNA. *Nature,* Vol.358, No.6385, (July 1992) pp. 422-424, ISSN 0028-0836

Buck, M.; Gallegos, M.T.; Studholme, D.J.; Guo, Y.; & Gralla, J.D. (2000). The Bacterial Enhancer-Dependent σ^{54} (σ^N) Transcription Factor. *Journal of Bacteriology,* Vol.182, No.15, (August 2000), pp. 4129-4136, ISSN 0021-9193

Bueno, S.M.; Riedel, C.A.; Carreño & Kalergis, A.M. (2010). Virulence Mechanisms Displayed by *Salmonella* to Impair Dendritic Cell Function. *Current Medical Chemistry,* Vol.17, No.12, (April 2010), pp. 1156-1166, ISSN 0929-8673

Burgess, R.R. & Travers, A.A. (1969). Factor Stimulating Transcription by RNA Polymerase. *Nature,* Vol.221, No.5175, (January 1969), pp. 43-46, ISSN 0028-0836

Burton, Z.F.; Gross, C.A.; Watanabe, K.K.; & Burgess, R.R (1983). The Operon That Encodes the Sigma Subunit of RNA Polymerase Also Encodes Ribosomal Protein S21 and DNA Primase in E. coli K12. *Cell,* Vol.32, No.2, (February 1983), pp. 335-349, ISSN 0092-8674

Cabeza, M.L.; Aguirre, A.; Soncini, F.C.; & Véscovi, E.G. (2007). Induction of RpoS Degradation by the Two-Component System Regulator RstA in *Salmonella enterica. Journal of Bacteriology,* Vol.189, No.20, (October 2007), pp. 7335-7342, ISSN 0021-9193

Chadsey, M.S.; Karlinsey, J.E.; & Hughes, K.T. (1998). The flagellar anti-σ factor FlgM actively dissociates *Salmonella* Typhimurium σ^{28} RNA polymerase holoenzyme. *Genes and Development,* Vol.12, No.19, (October 1998), pp. 3123–3136, ISSN 0890-9369

Chen, L.M.; Hobbie, S.; & Galán, J.E. (1996). Requirement of CDC42 of Salmonella-induced cytoskeletal and nuclear responses. *Science,* Vol.274, No.5295, (December 1996), pp. 2115-2118, ISSN 0036-8075

Chilcott, G.S. & Hughes, K.T. (2000). Coupling of Flagellar Gene Expression to Flagellar Assembly in *Salmonella enterica* Serovar Typhimurium and *Escherichia coli. Microbiology and Molecular Biology Review,* Vol.64, No.4, (December 2000), pp. 694-708, ISSN 1092-2172.

Cirillo, D.M.; Valdivia, R.H.; Monack, D.M.; & Falkow, S. (1998). Macrophage-dependent induction of the *Salmonella* pathogenicity island 2 type III secretion system and its role in intracellular survival. *Molecular Microbiology*, Vol.30, No.1, (October 1998), pp. 175-188, ISSN 0950-382X

Conlin, C.A. & Miller, C.G. (2000). *opdA*, a *Salmonella enterica* Serovar Typhimurium Gene Encoding a Protease, Is Part of an Operon Regulated by Heat Shock. *Journal of Bacteriology*, Vol.182, No.2, (January 2000), pp. 518-521, ISSN 0021-9193

Coornaert, A.; Lu, A.; Mandin, P.; Spring, M.; Gottesman, S.; & Guillier, M. (2010). MicA sRNA links the PhoP regulon to cell envelope stress. *Molecular Microbiology*, Vol.76, No.2, (April 2010), pp. 467-479, ISSN 0950-382X

Cróinín, T.Ó. & Dorman, C.J. (2007). Expression of the Fis protein is sustained in late-exponential- and stationary-phase cultures of *Salmonella enterica* serovar Typhimurium grown in the absence of aeration. *Molecular Microbiology*, Vol.66, No.1, (2007), pp. 237-251, ISSN 0950-382X

Cummings, L.A.; Rassoulian Barrett, S.L; Wilkerson, W.D.; Fellnerova, I.; & Cookson, B.T. (2005). FliC-Specific CD4+ T Cell Responses are Restricted by Bacterial Regulation of Antigent Expression. *Journal of Immunology*, Vol.174, No.12, (June 2005), pp. 7929-7938, ISSN 0022-1767

Cunning, C. & Elliot, T. (1999). RpoS Synthesis is Growth Rate Regulated in *Salmonella* Typhimurium, but Its Turnover Is Not Dependent on Acetyl Phosphate Synthesis on PTS Function. *Journal of Bacteriology*, Vol.181, No.16, (August 1999), pp. 4853-4862, ISSN 0021-9193

De Las Peñas, A.; Connolly, L.; & Gross, A.C. (1997) The σ^E –mediated response to extracytoplasmic stress in *Escherichia coli* is transduced by RseA and RseB, two negative regulators of σ^E . *Molecular Microbiology*, Vol.24, No.2, (April 1997), pp. 373-385, ISSN 0950-382X

Dodd, C.E.R. & Aldsworth, T.G. (2002). The importance of RpoS in the survival of bacteria through food processing. *International Journal of Food Microbiology*, Vol.74, No.3, (April 2002), pp. 189-194, ISSN 0168-1605

Dombroski, A.J.; Walter, W.A.; & Gross, C.A. (1993). Amino-terminal amino acids modulate sigma-factor DNA-binding activity. *Genes and Development*, Vol.7, No.12A, (December 1993), pp. 2446-2455, ISSN 0890-9369

Dombroski, A.J.; Walter, W.A.; Record, M.T.; Siegele, D.A.; & Gross, C.A. (1992). Polypeptides Containing Highly Conserved Regions of Transcription Initiation Factor σ^{70} Exhibit Specificity of Binding to Promoter DNA. *Cell*, Vol.70, No.3, (August 1992), pp. 501-512, ISSN 0092-8674

Dougan, G.; John, V.; Palmer, S.; & Mastroeni, P. (2011). Immunity to salmonellosis. *Immunology Reviews*, Vol.240, No.1, (March 2011), pp. 196-210, ISSN 0105-2896

Du, H.; Sheng, X.; Zhang, H.; Zou, X.; Ni, B.; Xu, S.; Zhu, Z.; Xu, H.; & Huang, X. (2011). RpoE may Promote Flagellar Gene Expression in *Salmonella enterica* Serovar Typhi Under Hyperosmotic Stress. *Current Microbiology*, Vol.62, No.2, (February 2011), pp.492-500, ISSN 0343-8651

Dunkley, K.D.; Callaway, T.R.; Chalova, V.I.; Anderson, R.C.; Kundinger, M.M.; Dunkley, C.S.; Nisbet, D.J.; & Ricke, S.C. (2008). Growth and genetic responses of *Salmonella* Typhimurium to pH-shifts in an anaerobic continuous culture. *Anaerobe*, Vol.14, No.1, (February 2008), pp. 35-42, ISSN 1075-9964

Eriksson, S.; Lucchini, S.; Thompson, A.; Rhen, M.; & Hinton, J.C.D. (2003). Unravelling the biology of macrophage infection by gene expression profiling of intracellular

Salmonella enterica. Molecular Microbiology, Vol.47, No.1, (2003), pp. 103-118, ISSN
0950-382X

Fang, F.C.; Libby, S.J.; Buchmeier, N.A.; Loewen, P.C.; Switala, J.; Harwood, J; & Guiney,
D.G. (1992). The alternative σ factor KatF (RpoS) regulates *Salmonella* virulence.
Proceedings of the National Acadmey of Sciences USA, Vol.89, No.24, (December 1992),
pp. 11978-11982, ISSN 0027-8424

Fang, F.C.; Chen, C.Y.; Guiney, D.G.; & Xu, Y. (1996). Identification of σ^S – Regulated Genes
in *Salmonella* Typhimurium: Complementary Regulatory Interactions between σ^S
and Cyclic AMP Receptor Protein. *Journal of Bacteriology*, Vol.178, No. 17,
(September 1996), pp. 5112-5120, ISSN 0021-9193

Fang, F.C. (2005). Sigma Cascades in Prokaryotic Regulatory Networks. *Proceedings of the
National Academy of Sciences*, Vol.102, (2005), pp. 4933-4934, ISSN 0027-8424

Farewell, A.; Kvint, K.; & Nyström, T. (1998) Negative regulation by RpoS: a case of sigma
factor competition. *Molecular Microbiology*, Vol.29, No.4, (August 1998), pp. 1039-
1051, ISSN 0950-382X

Francis, C.L.; Starnbach, M.N.; & Falkow, S. (1992). Morphological and cytoskeletal changes
in epithelial cells occur immediately upon interaction with *Salmonella*
Typhimurium grown under low-oxygen conditions. *Molecular Microbiology*, Vol.6,
No. 21, (November 1992), pp. 3077-3087, ISSN 0950-382X

Fu, Y. & Galán, J.E. (1999). A *Salmonella* protein antagonizes Rac-1 and Cdc42 to mediate
host-cell recovery after bacterial invasion. *Nature*, Vol.401, No.6750, (September
1999), pp. 293-297, ISSN 0028-0836

Galán, J.E. (1996). Molecular and Cellular Bases of *Salmonella* Entry into Host Cells. *Bacterial
Invasiveness*, Vol.209, (1996), pp. 43-60, ISSN 0070-217X

Guiney, D.G.; Libby, S.; Fang, F.C.; Krause, M.; & Fierer, J. (1995). Growth-phase regulation
of plasmid virulence genes in *Salmonella*. *Trends in Microbiology*, Vol.3, No.7, (July
1995), pp. 275-279, ISSN 0966-842X

Hansen-Wester, I. & Hensen, M. (2001). Salmonella pathogenicity islands encoding type III
secretion systems. *Microbes and Infection*, Vol.3, No.7, (June 2001), pp. 549-559, ISSN
1286-4579

Helmann, J.D. & Chamberlin, M.J. (1987). DNA sequence analysis suggests that expression
of flagellar and chemotaxis genes in *Escherichia coli* and *Salmonella Typhimurium* is
controlled by an alternate σ factor. *Proceedings of the National Academy of Sciences
USA*, Vol.84, No.18, (September 1987), pp. 6422-6424, ISSN 0027-8424

Hensel, M.; Shea, J.E.; Waterman, S.R.; Mundy, R.; Nikolaus, R.; Banks, G.; Vazquez-Torres,
A.; Gleeson, C.; Fang, F.C.; & Holden, D.W. (1998). Genes encoding putative
effector proteins of the type III secretion system of *Salmonella* pathogenicity island 2
are required for bacterial virulence and proliferation in macrophages. *Molecular
Microbiology*, Vol.30, No.1, (October 1998), pp. 163-174, ISSN 0950-382X

Hirsch, M. & Elliot, T. (2005). Fis Regulates Transcriptional Induction of RpoS in *Salmonella
enterica. Journal of Bacteriology*, Vol.187, No.5, (March 2005), pp. 1568-1580, ISSN
0021-9193

Hughes, K.T.; Gillen, K.L.; Semon, M.J.; & Karlinsey, J.E. (1993). Sensing Structural
Intermediates in Bacterial Flagellar Assembly by Export of a Negative Regulator.
Science, Vol.262, No.5137, (November 1993), pp. 1277-1280, ISSN 0036-8075

Ibanez-Ruiz, M.; Robbe-Saule, V.; Hermant, D.; Labrude, S.; and Norel, F. (2000). Identification of RpoS (σ^S)-regulated Genes in *Salmonella enterica* Serovar Typhimurium. *Journal of Bacteriology,* Vol.182, No.20, (October 2000), pp. 5749-5756, ISSN 0021-9193

Ikebe, T.; Iyoda, S.; & Kutsukake, K. (1999). Structure and expression of the *fliA* operon of *Salmonella* Typhimurium. *Microbiology,* Vol.145, No.6, (June 1999), pp. 1389-1396, ISSN 1350-0872

Ishihama, A. (1993). Protein-Protein Communication with the Transcription Apparatus. *Journal of Bacteriology,* Vol.175, No.9, (May 1993), pp. 2483-2489, ISSN 0021-9193

Jishage, M., & Ishihama,A. (1995). Regulation of RNA Polymerase Sigma Subunits Synthesis in *Escherichia coli:* Intracellular Levels of σ^{70} and σ^{38}. *Journal of Bacteriology,* Vol.177, No.23, (December 1995), pp. 6832-6835, ISSN 0021-9193

Jones, A.M.; Goodwill, A.; & Elliot, T. (2006). Limited Role for the DsrA and RprA Regulatory RNAs in *rpoS* Regulation in *Salmonella enterica. Journal of Bacteriology,* Vol.188, No.14, (July 2006), pp. 5077-5088, ISSN 0021-9193

Jones, B.D.; Ghori, N.; & Falkow, S. (1994). *Salmonella* Typhimurium Initiates Murine Infection by Penetrating and Destroying the Specialized Epithelial M Cells of the Peyer's Patches. *Journal of Experimental Medicine,* Vol.180, No.9, (Septemeber 1994), pp. 15-23, ISSN 0019-9567

Karlinsey, J.E.; Tanaka, S.; Bettenworth, V.; Yamaguchi, S.; Boos, W.; Aizawa, S.I.; & Hughes, K.T. (2000). Completion of the hook-basal body complex of the *Salmonella* Typhimurium flagellum is coupled to FlgM secretion and *fliC* transcription. *Molecular Microbiology,* Vol.37, No.5, (2000), pp. 1220-1231, ISSN 0950-382X

Karlinsey, J.E. & Hughes, K.T. (2006) Genetic Transplantation: *Salmonella enterica* Serovar Typhimurium as a Host to Study Sigma Factor and Anti-Sigma Factor Interactions in Genetically Intractable Systems. *Journal of Bacteriology,* Vol.188, No.1, (January 2006), pp. 103-114, ISSN 0021-9193

Kenyon, W. J.; Thomas, S.M.; Johson, E.; Pallen, M.J.; & Spector, M.P. (2005). Shifts from glucose to certain secondary carbon-sources result in activation of the extracytoplasmic function sigma factor σ^E in *Salmonella enterica* serovar Typhimurium. *Microbiology,* Vol.151, No.7, (July 2005), pp. 2373-2383, ISSN 1350-0872

Khan, A.Q.; Zhao, L.; Hirose, K.; Miyake, M.; Li, T.; Hashimoto, Y.; Kawamura, Y.; & Ezaki, T. (1998). *Salmonella typhi rpoS* mutant is less cytotoxic than the parent strain but survives inside resting THP-1 macrophages. *FEMS Microbiology Letters,* Vol.161, No.1, (April 1998), pp. 201-208, ISSN 0378-1097

Kilstrup, M.; Jessing, S.G.; Wichmand-Jørgensen, S.B.; Madsen, M.; & Nilsson, D. (1998). Activation Control of *pur* Gene Expression in *Lactococcus lactis*a; Proposal for a Consensus Activator Binding Sequence Based on Deletion Analysis and Site-Directed Mutagenesis of *purC* and *purD* Promoter Regions. *Journal of Bacteriology,* Vol.180, No.15, (August 1998), pp. 3900-3906, ISSN 0021-9193

Komitopoulou, E.; Bainton, N.J.; & Adams, M.R. (2004). Oxidation-reduction potential regulates RpoS levels in *Salmonella* Typhimurium. *Journal of Applied Microbiology,* Vol.96, No.2, (February 2004), pp. 271-278, ISSN 1384-5072

Kowarz, L.; Coynault, C.; Robbe-Saule, V.; & Norel, F. (1994). The *Salmonella* Typhimurium *katF* (*rpoS*) Gene: Cloning, Nucleotide Sequence, and Regulation of *spvR* and *spvABCD* Virulence Plasmid Genes. *Journal of Bacteriology*, Vol.176, No.22, (November 1994), pp. 6852-6860, ISSN 0021-9193

Kutsukake, K. & Iino, T. 1994. Role of the FliA-FlgM Regulatory System on Transcriptional Control of the Flagellar Regulon and Flagellar Formation in *Salmonella* Typhimurium. *Journal of Bacteriology*, Vol.176, No.12, (June 1994), pp. 3598-3605, ISSN 0021-9193

Kutsukake, K.; Iyoda, S.; Ohnishi, K.; & Iino, T. (1994). Genetic and molecular analysis of the interaction between the flagellum-specific and anti-sigma factors in *Salmonella* Typhimurium. *Journal of Environmental Microbiology*, Vol.13, No.19, (October 1994), pp. 4568-4576, ISSN 0261-4189

Lara-Tejero, M. & Galán, J.E. (2009). *Salmonella enterica* Serovar Typhimurium Pathogenicity Island 1-Encoded Type III Secretion System Translocases Mediate Intimate Attachement to Nonphagocytic Cells. *Infection and Immunology*, Vol.77, No.1, (July 2009), pp. 2635-2642, ISSN 0019-9567

Lawhon, S.D.; Maurer, R.; Suyemoto, M.; & Altier, C. (2002). Intestinal short-chain fatty acids alter *Salmonella* Typhimurium invasion gene expression and virulence though BarA/SirA. *Molecular Microbiology*, Vol.46, No.5 (2002), pp. 1451-1464, ISSN 0950-382X

Lee, H.Y.; Cho, S.A.; Lee, I.S.; Park, J.H.; Seok, S.H.; Bae, M.W.; Kim, D.J.; Lee, S.H.; Hur, S.J.; Ban, S.J.; Lee, Y.K.; Han, Y.K.; Cho, Y.K.; & Park, J.H. (2007). Evaluation of *phoP* and *rpoS* mutants of *Salmonella enterica* serovar Typhi as attenuated vaccine candidates: virulence and protective immune responses in intranasally immunized mice. *FEMS Immunology Med Microbiol*, Vol.51, No.2, (2007), pp. 310-318, ISSN 0928-8244

Liu, X. & Matsumura, P. (1994). The FlhD/FlhC Complex, a Transcriptional Activator of the *Escherichia coli* Flagellar Class II Operons. *Journal of Bacteriology*, Vol.176, No.23, (December 1994), pp. 7345-7351, ISSN 0021-9193

Liu, X.; Fukita, N.; Ishihama, A.; & Matsumara, P. (1995). The C-Terminal region of the α Subunit of *Escherichia coli* RNA Polymerase is Required for Transcriptional Activation of the Flagellar Level II Operons by the FlhD/FlhC Complex. *Journal of Bacteriology*, Vol.177, No.17, (September 1995), pp. 5186-5188, ISSN 0021-9193

Löber, S.; Jäckel, D.; Kaiser, N.; & Hensel, M. (2006). Regulation of *Salmonella* pathogenicity island 2 genes by independent environmental signals. *International Journal of Medical Microbiology*, Vol.296, No.7, (Novemeber 2006), pp. 435-447, ISSN 1438-4221

Lucas, R.L.; Lostroh, C.P.; DiRusso, C.C.; Spector, M.P.; Wanner, B.L.; & Lee, C.A. (2000). Multiple Factors Independently Regulate *hila* and Invasion Gene Expression in *Salmonella enterica* Serovar Typhimurium. *Journal of Bacteriology*, Vol.182, No.7, (April 2000), pp. 1872-1882, ISSN 0021-9193

Lucchini, S.; McDermott, P.; Thompson, A.; & Hinton, J.C.D. (2009). The H-NS-like protein StpA represses the RpoS (σ^{38}) regulon during exponential growth of *Salmonella* Typhimurium. *Molecular Microbiology*, Vol.74, No.5, (December 2009), pp. 1169-1186, ISSN 0950-382X

Majdalani, N.; Chen, S.; Murron, J.; St John, K.; & Gottesman, S. (2001). Regulation of RpoS by a novel small RNA: the characterization of RprA. *Molecular Microbiology*, Vol.39, No.5, (March 2001), pp. 1382-1394, ISSN 0950-382X

Mallik, P.; Pratt, T.S.; Beach, M.B.; Bradley, M.D.; Undamatla, J.; & Osuna, R. (2004). Growth Phase-Dependent Regulation and Stringent Control of *fis* are Conserved Processes in Enteric Bacteria and Involve a Single Promoter (*fis* P) in *Escherichia coli*. *Journal of Bacteriology*, Vol.186, No.1, (January 2004), pp. 122-135, ISSN 0021-9193

Matsui, M.; Takaya, A.; & Yamamoto, T. (2008). σ^{38} – Mediated Negative Regulation of *Salmonella* Pathogenicity Island 1 Expression. *Journal of Bacteriology*, Vol.190, No.20, (October 2008), pp. 6636-6645, ISSN 0021-9193

McDermott, J.E.; Yoon, H.; Nakayasu, E.S.; Metz, T.O.; Hyduke, D.R.; Kidawi, A.S.; Palsson, B.O.; Adkins, J.N.; & Heffron, F. (2011). Technologies and approaches to elucidate and model the virulence program of *Salmonella*. *Frontiers in Microbiology*, Vol.2, No.121, (June 2011), pp.1-14., ISSN 1664-302X

McMeechan, A.; Robers, M.; Cogan, T.A.; Jørgensen, F.; Stevenson, A.; Lewis, C.; Rowley, G.; & Humphrey, T.J. (2007). Role of the alternative sigma factors σ^E and σ^S in survival of *Salmonella enterica* serovar Typhimurium during starvation, refrigeration, and osmotic shock. *Microbiology*, Vol.153, No.1, (January 2007), pp. 263-269, ISSN 1350-0872

Minagawa, S.; Ogasawara, H.; Kato, A.; Yamamoto, K.; Eguchi, Y.; Oshima, T.; Mori, H.; Ishihama, A.; & Utsumi, R. (2003). Identification and Molecular Characterization of the Mg^{2+} Stimulon of *Escherichia coli*. *Journal of Bacteriology*, Vol.185, No.13, (July 2003), pp. 3696-3702, ISSN 0021-9193

Monsieurs, P.; De keersmaecker, S.; Navarre, W.W.; Bader, M.W.; De Smet, F.; McClelland, M.; Fang, F.C.; De Moor, B.; Vanderleyden, J.; & Marchal, K. (2005) Comparison of the PhoPQ Regulon in *Escherichia coli* and *Salmonella* Typhimurium. Vol. 60, No. 4 (April 2005), pp. 462-474, ISSN 0022-2844.

Morett, E. & Segovia, L. (1993). The σ^{54} Bacterial Enhancer-Binding Protein Family: Mechanism of Action and Phylogenetic Relationship of Their Functional Domains. *Journal of Bacteriology*, Vol.175, No.19, (October 1993), pp. 6067-6074, ISSN 0021-9193

Muller, C.; Bang, I.S.; Velayudhan, J.; Karlinsey, J.; Paperfort, K.; Vogel, J.; & Fang, F.C. (2009). Acid stress activation of the σ^E stress response in *Salmonella enterica* serovar Typhimurium. *Molecular Microbiology*, Vol.71, No.5, (2009), pp. 1228-1238, ISSN 0950-382X

Mutalik, V.K.: Nonaka, G.; Ades, S.E.; Rhodius, V.A.; & Gross, C.A. (2009). Promoter Strength Properties of the Complete Sigma E Regulon on *Escherichia coli* and *Salmonella enterica*. *Journal of Bacteriology*, Vol.191, No.23 (December 2009), pp. 7279-7287, ISSN 0021-9193

O'Neal, C.R.; Gabriel, W.M.; Turk, A.K.; Libby, S.J.; Fang, F.C.; & Spector, M. (1994). RpoS Is Necessary for Both Positive and Negative Regulation of Starvation Survival Genes during Phosphate, Carbon, and Nitrogen Starvation in *Salmonella* Typhimurium. *Journal of Bacteriology*, Vol.176, No.15, (August 1994), pp. 4610-4616, ISSN 0021-9193

Ohl, M.E. & Miller, S.I. (2001). *Salmonella*: A Model for Bacteria Pathogenesis. *Annual Review of Medicine*, Vol.52, (2001), pp. 259-274, ISSN 0066-4219.

Ohnishi, K.; Katsukake, K.; Suzuki, H.; & Iino, T. (1990). Gene *fliA* encodes an alternative sigma factor specific for flagellar operons in *Salmonella* Typhimurium. *Molecular and General Genetics*, Vol.221, No.2, (April 1990), pp. 139-147, ISSN 0026-8925

Osborne, S.E. & Coombes, B.K. (2009). RpoE fine tunes expression of a subset of SsrB-regulated virulence factors in *Salmonella enterica* serovar Typhimurium. *BMC Microbiololgy*, Vol.9, No.1, (March 2009), pp. 1-10, ISSN 1471-2180

Papenfort, K.; Pfeiffer, V.; Mika, F.; Lucchini, S.; Hinton, J.C.D.; & Vogel, J. (2006). σ^E – dependent small RNAs of *Salmonella* respond to membrane stress of accelerating global omp mRNA decay. *Molecular Microbiology*, Vol.62, No.6, (2006), pp. 1674-1688, ISSN 0950-382X

Rakjumari, K. & Gowrishankar, J. (2001). In Vivo Expression from the RpoS-Dependent P1 Promoter of the Osmotically Regulated *proU* Operon in *Escherichia coli* and *Salmonella enterica* Serovar Typhimurium: Acivation by *rho* and *hns* Mutations and by Cold Stress. *Journal of Bacteriology*, Vol.183, No.22, (November 2001), pp. 6543-6550, ISSN 0021-9193

Rhodius, V.A.; Suh, W.C.; Nonaka, G.; West, J, & Gross, C.A. (2006) Conserved and variable functions of the σ^E-stress response in related genomes. *PLoS Biology*, Vol.4, No.1, (January 2006), pp. e2, ISSN 1544-9173

Robbe-Saule, V.; Lopes, M.D.; Kolb, A.; & Norel, F. (2007). Physiological Effects of Crl in *Salmonella* Are Modulated by σ^S Level and Promoter Specificity. *Journal of Bacteriology*, Vol.189, No.8, (April 2007), pp. 2976-2987, ISSN 0021-9193

Römling, U.; Sierralta, W.D.; Eriksson, K., & Normark, S. (1998). Multicellular and aggregative behavior of *Salmonella* Typhimurium strains is controlled by mutations in the *agfD* promoter. *Molecular Microbiology*, Vol.28, No.2, (April 1998), pp. 249-264, ISSN 0950-382X

Scaife, J.G.; Heilig, J.S.; Rowen, L.; & Calendar, R. (1979). Gene for the RNA polymerase σ subunit mapped in *Salmonella* Typhimurium and *Escherichia coli* by cloning and deletion. *Proceedings of the National Academy of Sciences USA*, Vol.76, No.12, (December 1979), pp. 6510-6514, ISSN 0027-8424

Shea, J.E.; Hensel, M.; Gleeson, C.; & Holden, D.W. (1996). Identification of a virulence locus encoding a secondy type III secretion system in *Salmonella* Typhimurium. *Proceedings of the National Academy of Sciences USA*, Vol.93, No.6, (March 1996), pp. 2593-2597, ISSN 0027-8424

Sittka, A.; Pfeiffer, V.; Tedin, K; & Vogel, J. (2007). The RNA chaperone Hfq is essential for the virulence of *Salmonella* Typhimurium. *Molecular Microbiology*, Vol.63, No.1, (2007), pp. 193-217, ISSN 0950-382X

Slauch, J.; Taylor, R.; & Maloy, S. (1997). Survival in a cruel world: how *Vibrio cholera* and *Salmonella* respond to an unwilling host. *Genes and Development*, Vol.11, No.14, (July 1997), pp. 1761-1774, 0890-9369

Steele-Mortimer, O.; Brumell, J.H.; Knodler, L.A.; Méresse, S.; Lopez, A.; & Finlay, B.B. (2002). The invasion-associated type III associated secretion system of *Salmonella enterica* serovar Typhimurium is necessary for intracellular proliferation and vacuole biogenesis in epithelial cells. *Cell Microbiology*, Vol.4, No.1, (January 2002), pp. 43-54, ISSN 1462-5814

Studholme, D.J. (2002). Enhancer-Dependent Trancription in *Salmonella enterica* Typhimurium: New Members of the σ^N Regulon Inferred from Protein Sequence Homology and Predicted Promoter Sites. *Journal of Molecular Microbiology and Biotechnology*, Vol.4, No.4, (July 2002), pp. 367-374, ISSN 1462-5814

Sutton, A.; Buencamino, R.; & Eisenstark, A. (2000). *rpoS* Mutants in Archival Cultures of *Salmonella enterica* Serovar Typhimurium. *Journal of Bacteriology*, Vol.182, No.16, (August 2000), pp. 4375-4379, ISSN 0021-9193

Testerman, T.L.; Vasquez-Torres, A.; Xu, T.; Jones-Carson, J.; Libby, S.J.; & Fang, F.C. (2002). The alternative sigma factor σ^E controls antioxidant defences required for *Salmonella* virulence and stationary-phase survival. *Molecular Microbiology,* Vol.43, No.5, (2002), pp. 771-782, ISSN 0950-382X

Tobar, J.A.; Carreño, L.J.; Bueno, S.M.; González, P.A.; Mora, J.E.; Quezada, S.A.; & Kalergis, A.M. (2006). Virulent *Salmonella enterica* Serovar Typhimurium Evades Adaptive Immunity by Preventing Dendritic Cells from Activating T Cells. *Infection and Immunity,* Vol.74, No.11, (November 2006), pp. 6438-6448, ISSN 0019-9567

Tu, X.; Latifi, T.; Bougdour, A.; Gottesman, S.; & Groisman, E.A. (2006). The PhoP/PhoQ two-component system stabilizes the alternative sigma factor RpoS in *Salmonella enterica. Proceedings of the National Academy of Sciences,* Vol.103, No.36, (September 2006), pp. 13503-13508, ISSN 0027-8424

Waldminghaus, T.; Heldrick, J.; Brantl, S.; & Narberhaus, F. (2007). FourU: a novel type of RNA thermometer in *Salmonella. Molecular Microbiology,* Vol.65, No.2, (2007), pp. 413-424, ISSN 0950-382X

Walker, K.A.; Atkins, C.L.; & Osuna, R. (1999). Functional Determinants of the *Escherichia coli fis* Promoter: Role of -35, -10, and Transcription Initiation Regions in the Response to Stringent Control and Growth Phase-Dependent Regulation. *Journal of Bacteriology,* Vol.181, No.4, (February 1999), pp. 1269-1280, ISSN 0021-9193

Walker, K.A.; Mallik, P.; Pratt, T.S.; & Osuna, R. (2004). The *Escherichia coli fis* Promoter is Regulated by Changes in the Levels of Its Transcription Initiation Nucleotide CTP. *Journal of Biological Chemistry,* Vol.279, No.49, (December 2004), pp. 50818-50828, ISSN 0021-9258

Webb, C.; Moreno, M.; Wilmes-Riesenberg, M.; Curtis III, R.; & Foster, J.W. (1999). Effects of DksA and ClpP protease on sigma S production and virulence in *Salmonella* Typhimurium. *Molecular Microbiology,* Vol.34, No.1, (October 1999), pp. 112-123, ISSN 0950-382X

Wood, L.F.; Leech, A.J.; & Ohman, D.E. (2006). Cell wall-inhibitory antibiotics activate the alginate biosynthesis operon in *Pseudomonas aeruginosa*: roles of σ^{22} (AlgT) and the AlgW and Prc proteases. *Molecular Microbiology,* Vol.62, No.2, (2006), pp. 412-426, ISSN 0950-382X

Wozniak, C.E.; Chevance, F.F.V.; & Hughes, K.T. (2010). Multiple Promoters Contribute to Swarming and the Coordination of Transcription with Flagellar Assembly in *Salmonella. Journal of Bacteriology,* Vol.192, No.18, (September 2010), pp. 4752-4762, ISSN 0021-9193

Yokoseki, T.; Iino, T.; & Kutsukake, K. (1996). Negative Regulation of FliD, FliS, and FliT of the Export of the Flagellum-Specfic Anti-Sigma Factor, FlgM, in *Salmonella* Typhimurium. *Journal of Bacteriology,* Vol.178, No.3, (February 1996), pp. 899-901, ISSN 0021-9193

6

Molecular Armory of *S.* Typhi: Deciphering the Putative Arsenal of Our Enemy

Chantal G. Forest and France Daigle
Department of Microbiology & Immunology/University of Montreal
Canada

1. Introduction

The outer surface of bacteria is the first to interact with host components, such as the immune system, the extracellular matrix or cells. The bacterial Gram-negative cell wall is complex and composed of an inner membrane (IM), a periplasmic space and a thin peptidoglycan layer, all surrounded by an outer membrane (OM). The OM is a bilayered structure consisting mainly of phospholipids, proteins and lipopolysaccharide (LPS) and serves as an impermeable barrier to prevent the escape of periplasmic molecules but also acts as a barrier for entry of external molecules. *Salmonella enterica* comprises more than 2500 serovars, based on three major antigens located at the cell surface: O antigen, capsule and flagella. All serovars are highly conserved genetically but have different host ranges and cause different diseases. In humans, *Salmonella* infection causes gastroenteritis, often associated with serovars Typhimurium and Enteritidis or typhoid-like disease, which is associated with serovars Typhi and Paratyphi. *S.* Typhi strains belong to serogroup D1 with the antigenic formula O:9,12; Vi+; H-d. These strains are human-restricted and besides asymptomatic carriers, no environmental reservoir is known.

S. Typhi is a monomorphic bacterium, showing very little genetic diversity (Kidgell et al., 2002) and up to 5% of its annotated coding sequences are pseudogenes (Holt et al., 2009; Parkhill et al., 2001). Genome degradation may be responsible for its host specificity; however the *S.* Typhi genome may harbour specific genes for its systemic dispersion and survival. *S.* Typhi remains a major public health problem in developing countries. Antimicrobial resistance has become a problem in endemic regions, and it is becoming imperative to develop new vaccine strategies or discover new antimicrobial targets to combat this microorganism. Bacterial surface proteins may correspond to these targets by being immunogenic or essential for virulence. Most virulence factors are usually located within genomic locations called *Salmonella* Pathogenicity Islands (SPIs) and are tightly regulated by global regulators such as PhoP-PhoQ, RcsDBC, OmpR-EnvZ and RpoS. This review will focus on molecules localized at the outer membrane of *S.* Typhi and their role in pathogenesis. A complete analysis of adhesive molecules, such as the 12 fimbrial systems, curli, type IVB pilus, autotransporters and afimbrial adhesins will be presented. We will also discuss the importance of polysaccharides such as the Vi capsule and LPS. Furthermore, the complex surface structures generated by secretion systems, such as type three secretion systems (T3SS), flagella and T6SS that are so important for invasion, intracellular survival and to highjack the host defence system will be discussed. Finally, methods used to inhibit these adhesive structures will be described.

2. Fimbrial adhesins

Fimbriae (also called pili) are proteinaceous structures that can be observed as filaments anchored on the bacterial cell surface. These structures can mediate crucial interactions during host infection like adherence, invasion or biofilm formation, and are classified according to their mechanism of assembly. Most of the fimbriae present in S. Typhi genome are assembled by the chaperone/usher pathway, but there are also one representative of the nucleation/precipitation pathway (*csg*) and one type IVB pilus. This section will briefly describe each mechanism of expression and the current knowledge related to S. Typhi and their putative roles.

2.1 Mechanisms of fimbrial assembly

2.1.1 The chaperone/usher pathway

Twelve fimbrial systems detected in S. Typhi belong to the chaperone/usher pathway (CUP) assembly class (Fig. 1). A classic fimbrial operon usually harbours at least four different genes. The filaments are composed of major and minor fimbrial subunits assembled by the cooperative work of the chaperone and the usher. After translocation by

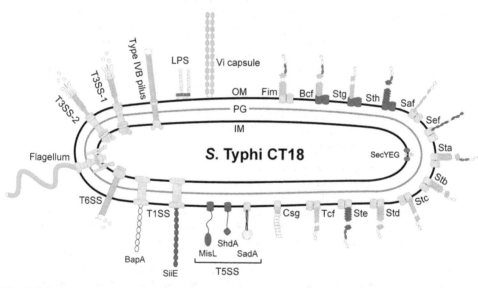

Fig. 1. Schematic representation of the important surface structures identified in S. Typhi CT18. Adhesive components are shown in yellow, membrane components are cyan and pseudogenes are shown in blue (pseudogenes of the T6SS are not shown). The twelve representatives of the CUP are grouped in fimbrial clades and are drawn according to previous observations (Salih et al., 2008) or based on their homologues found in E. coli K-12 (Korea et al., 2011). STY0405 putative autotransporter, STY0351 adhesin, and PagC which are known to be implicated in virulence were omitted from the drawing, as well as STY1980 (MAM7). IM stands for inner membrane, PG for peptidoglycan and OM for outer membrane.

the Sec general secretory pathway, the periplasmic chaperone protects the subunits and brings them to the OM usher, which specifically translocates subunits to the cell surface. Fimbrial biogenesis by the CUP pathway is a self-energized process catalyzed by both the usher and the presence of high-energy intermediates in the folding of the chaperone-subunit complexes (Jacob-Dubuisson et al., 1994; Nishiyama et al., 2008; Sauer et al., 2002; Zavialov et al., 2002). Classification based upon sequence homology between the different ushers (Nuccio & Bäumler, 2007) revealed members in the γ1- (*bcf, fim, stg, sth*), γ3- (*saf, sef*), γ4- (*sta, stb, stc*), π- (*std, ste*) and the α-fimbrial clades (*tcf*) in S. Typhi (Fig. 1).

2.1.2 Nucleation/precipitation pathway

The thin aggregative fimbriae, also known as curli or TAFI, encoded by the *csgDEFG csgABC* gene cluster belongs to this class of adhesin. The first steps of biogenesis are similar to the CUP : after translocation by the Sec pathway, CsgA and CsgB fimbrial subunits are secreted by the CsgG outer membrane protein at the bacterial cell surface. The major difference between curli and CUP lies in its extracellular fiber growth assembly (Hammar et al., 1996). After secretion of the CsgB subunit, CsgA precipitates, polymerizes on CsgB and adopts an insoluble structure related to amyloid fibers (Hammar et al., 1996).

2.1.3 Type IVB pili

One of the most studied adhesive structures of S. Typhi is the type IVB pilus encoded by the *pil* operon located on SPI-7. Although type IV pili also produce long and flexible structures on the bacterial cell surface, their mechanism of assembly strongly differs from the CUP and curli pathways as it requires many structural proteins and is an ATP-dependent process. First, PilS prepilins are translocated through the IM into the periplasm and a specific prepilin peptidase cleaves the N-terminal signal peptide (reviewed in Craig & Li, 2008). An integral IM protein mobilizes a specific ATPase from the cytoplasm which drives pilus assembly. An oligomeric channel called the secretin found in the OM allows the exit of the pilus at the cell surface of the bacteria. ATP hydrolysis moves the pilus out in the secretin pore allowing the recruitment of new prepilin subunits. Unlike CUP and Csg fimbriae, Type IV pili are still connected to the IM of the bacteria and can be retracted rapidly inside the bacteria.

2.2 Roles of fimbrial adhesins during typhoid fever

In most studies, *Salmonella* fimbriae are involved during intestinal colonization (Althouse et al., 2003; Chessa et al., 2009; Weening et al., 2005), or in biofilm formation (Boddicker et al., 2002; Ledeboer et al., 2006), although they can also be used during the systemic phase (Edwards et al., 2000; Lawley et al., 2006). Interestingly, each serovar of *Salmonella enterica* harbours a unique combination of fimbrial operons, probably to avoid cross-immunity between two serovars infecting the same host (Norris & Bäumler, 1999; Nuccio et al., 2011). As S. Typhi infects only humans, little is known regarding the conditions of expression or the implication of each fimbrial adhesin during the course of infection. While some clues may be found in the literature, there is still much work to be done. Three fimbrial systems are clustered within pathogenic islands: *tcf* (Typhi colonizing factor) and *saf* (*Salmonella* atypical fimbriae) are found within SPI-6, while *sef* is in SPI-10 (Sabbagh et al., 2010).

Proteins expressed during infection were detected in blood of patients with typhoid fever (Charles et al., 2010; Harris et al., 2006; Hu et al., 2009). Interestingly, six proteins related to fimbrial adhesins led to the formation of antibodies after typhoid fever (TcfB, StbD, CsgG, CsgF, CsgE and BcfD). Since three proteins belonging to the thin aggregative fimbriae were identified, it suggests a strong production *in vivo* as well as an important role during infection. Csg implication during attachment to surfaces, bacterial autoaggregation and in biofilm formation is well known for *S.* Typhimurium and *E. coli* (Jonas et al., 2007). Nevertheless, a clear characterization of *csg* is needed for *S.* Typhi as there seem to be variations in expression between the different isolates (Römling et al., 2003; White et al., 2006). In *S.* Typhi, a strong expression of *csg* and *saf* fimbrial operons was observed inside human macrophages (Faucher et al., 2006).

The *bcf*, *sef*, *ste*, *stg* and *sth* fimbrial systems harbour pseudogenes that might disrupt the production of the corresponding fimbriae (Townsend et al., 2001). However, deletion of *stg* leads to reduced adhesion on epithelial cells as well as enhancement of the phagocytosis rate by macrophages (Forest et al., 2007). Furthermore, the presence of antibodies directed against BcfD is intriguing since the *bcfC* usher harbours two premature stop codons (Parkhill et al., 2001). The Bcf, Stb, Stc, Std and Sth fimbrial systems are required for the intestinal persistence of *S.* Typhimurium in mice, but their roles during the pathogenesis of *S.* Typhi still need to be evaluated (Weening et al., 2005). Sta and Tcf do not seem to be used for adhesion or invasion of non-polarized human epithelial cells while both are expressed at high NaCl concentrations (Bishop et al., 2008). Since these two fimbriae are found almost exclusively in the genome of serovars causing typhoid fever, they might be involved during the systemic phase or for the chronic carrier state (Nuccio et al., 2011). Although roles for Saf (Carnell et al., 2007; Lawley et al., 2006), Sef (Edwards et al., 2000) and Std fimbriae (Chessa et al., 2008; Weening et al., 2005) have been observed in other serovars of *Salmonella*, their true implication during typhoid fever needs to be investigated.

Type 1 fimbriae encoded by the *fim* operon are the best studied fimbrial adhesins and are frequently found in enteric bacteria. Fim are characterized by their mannose-sensitive binding properties, but their cell tropism seems to vary greatly between species and even between different strains of the same serovar (Thankavel et al., 1999). In *S.* Typhi, most clinical strains are fimbriated (*fim*+) and afimbriated strains are less adhesive and invasive than the fimbriated ones (Duguid et al., 1966; Satta et al., 1993). The ability of type 1 fimbriae to agglutinate yeast is abolished when the Vi capsule is expressed (Miyake et al., 1998). In *S.* Typhimurium, Fim appears to be the only fimbrial adhesin expressed in Luria-Bertani (LB) broth as confirmed by electron microscopy and flow cytometry (Duguid et al., 1966; Humphries et al., 2003). In *S.* Typhi, a complete deletion of *fim* also showed no evident fimbrial structures on the cell surface of the bacteria after growth in LB broth (Fig. 2).

Type IVB pili interact with the cystic fibrosis transmembrane conductance regulator (CFTR), a receptor upregulated and actively used by *S.* Typhi for its interaction with human epithelial cells (Lyczak & Pier, 2002; Pier et al., 1998; Tsui et al., 2003). These pili can also mediate bacterial self-association in conditions found in the intestinal tract, probably by enhancing binding efficiency prior to cell invasion (Morris et al., 2003a; Morris et al., 2003b). A direct correlation was observed between the level of surface-exposed CFTR and the efficiency of invasion of *S.* Typhi through the intestinal barrier (Pier et al., 1998). This specific interaction can be blocked by the addition of prepilin pre-PilS in the cell culture medium or with monoclonal antibodies specific to the first extracellular domain of CFTR

(Pier et al., 1998; Zhang et al., 2000). A piliated strain also adheres and invades human monocytes in a greater extent than a non-piliated strain and its expression can also increase IL-6 and NF-kappa B production in human monocytes by activating protein kinase C (Pan et al., 2005; F. Wang et al., 2005). Only a few other serovars, such as S. Paratyphi B and C, S. Heidelberg and S. *bongori* possess the genetic information coding for type IVB pili (Nuccio et al., 2011). Other functions could potentially be found in future studies as Type IV pili are also implicated in a variety of processes like biofilm formation, immune escape, DNA uptake and phage transduction in other pathogenic bacteria (reviewed in Craig & Li, 2008). These pili can also act as pistons, retracting subunits into the bacteria while it is still attached to a surface in a mechanism called "twitching motility" providing flagella-independent motility (reviewed in Mattick, 2002).

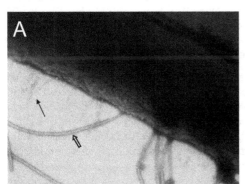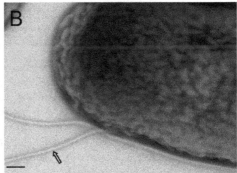

Fig. 2. Surface observation of S. Typhi grown in LB broth by transmission electron microscopy. After negative staining with phosphotungstate 1%, fimbriae were observed at the cell surface of the wild-type ISP1820 strain (A), while no structure was observed when *fim* was deleted (B). Black arrow shows fimbria and the open arrows indicate flagella. Black bar = 100 nm.

No genes related to fimbrial operons were found after a screening for mutants with a competitive disadvantage in humanized mice engrafted with hematopoietic stem cells (Libby et al., 2010). This result strongly suggests that fimbrial operons are mostly required during interaction with human epithelial cells, such as intestinal and gallbladder cells, that are absent from this mice model. Moreover, functional redundancy is often observed for fimbrial adhesins making it hard to evaluate their true contribution by single mutations. In order to understand the specific role played by each fimbrial system, our laboratory is currently creating a S. Typhi strain with deletions of all its fimbrial adhesins. This strain will greatly help to evaluate the global contribution of each fimbrial adhesins during association with eukaryotic cells.

3. Non fimbrial adhesins

3.1 Type 1 secretion systems

In *Salmonella*, some important surface structures are expressed by different mechanisms and can be classified as non fimbrial adhesins. In S. Typhi, there are two examples of adhesins secreted by a type I secretion system (T1SS) : SiiE and BapA. In T1SS, the secreted proteins directly pass through a channel formed between the IM and OM of the bacteria by the recognition of a signal at the C-terminus (China & Goffaux, 1999 as cited in Main-Hester et al., 2008). SPI-4, present in all *Salmonella* strains, encodes a T1SS responsible for the secretion of SiiE, the largest protein found in *Salmonella* (595 kDa) (Latasa et al., 2005; Main-Hester et al., 2008). Its cell surface expression requires the IM ATPase SiiF, the periplasmic adaptor SiiD and an outer membrane channel formed by SiiC (Gerlach et al., 2007). This adhesion system acts in a coordinated way with the T3SS of SPI-1 and is involved during the intestinal phase of infection (Gerlach et al., 2008). Previously annotated as two distinct ORFs in *S.* Typhi (STY4458-4459) (Parkhill et al., 2001), *siiE* harbours a premature stop codon probably rendering this large adhesin non-functional (Main-Hester et al., 2008; Morgan et al., 2004). An immunoblot done with antibodies directed against SiiE (STY4458) demonstrated the absence of production of SiiE in the whole cell proteins of *S.* Typhi further suggesting a lack of function in this serovar (Main-Hester et al., 2008). However, a transposon insertion in STY4458 showed a reduced competitive fitness in humanized mice, suggesting SiiE functionality and an uncharacterized role during interaction with hematopoietic cells (Libby et al., 2010).

A second T1SS is clustered within SPI-9 and secretes another large repetitive protein called BapA (biofilm-associated protein) due to its similarity with BapA of *Staphylococcus aureus*. Well described in *S.* Enteritidis, BapA is involved in bacterial autoaggregation strongly inducing biofilm formation and is also required during the interaction with the intestinal mucosa (Latasa et al., 2005). Its expression is under the control of CsgD, an important regulator also coordinating curli fimbriae and cellulose production needed for biofilm production (Jonas et al., 2007). Again, solving the components required for biofilm formation by *S.* Typhi is crucial since *bcsC* (STY4184), essential for cellulose and biofilm production, is a pseudogene (Parkhill et al., 2001; Zogaj et al., 2001).

3.2 Type 5 secretion systems

Autotransported adhesins can be monomeric or trimeric and are considered as type 5 secretion systems. *S.* Typhi harbours two monomeric examples of autotransported adhesins, *shdA* (STY2755) found in the CS54 island and *misL* (STY4030) clustered in SPI-3, as well as one representative of a trimeric autotransporter called *sadA* (STY4105). An N-terminal signal sequence allows their translocation into the periplasm by the Sec general secretory pathway, then a β-domain found at the C-terminal end of the protein adopts a β-barrel conformation in the OM allowing secretion of the passenger domain into the extracellular space (reviewed in Nishimura et al., 2010). ShdA is widely distributed in *S. enterica* subspecies I and appears to be produced during typhoid fever despite the presence of a frameshifting sequence (Harris et al., 2006; Parkhill et al., 2001). Interestingly, ShdA and MisL can bind fibronectin in other serovars of *Salmonella* and are both considered as pseudogenes in *S.* Typhi (Dorsey et al., 2005; Kingsley et al., 2002). SadA harbours homology to the trimeric autotransporter adhesin YadA of *Yersinia enterocolitica*, a highly repetitive fibrous surface protein (Grosskinsky et al., 2007). YaiU (STY0405) encodes a fourth putative autotransported adhesin with no known role except that antibodies against the protein are produced during a typhoid fever (Harris et al., 2006).

3.3 Other adhesins

Besides fimbrial and afimbrial adhesins, other surface-exposed proteins can act as adhesins and mediate crucial roles during typhoid fever. One of the most hydrophobic proteins encoded in the S. Typhi chromosome, STY0351, was recently characterized in detail and might be used as a potential vaccine target. This cell-surface protein is a novel adhesin directly involved in the pathogenesis of S. Typhi by conferring strong binding to the laminin extracellular matrix (Ghosh et al., 2011) and is positively regulated by the PhoP-PhoQ two-component system (Charles et al., 2009). It also possesses high immunogenic properties and STY0351-specific antibodies confer protection in a mouse model (Charles et al., 2010; Ghosh et al., 2011). PagC is another surface-exposed protein activated by the PhoP-PhoQ system that is produced and actively recognised by antibodies from patients having previously suffered from typhoid fever (Charles et al., 2010; Harris et al., 2006). Previously associated with survival within macrophages (Miller et al., 1989), PagC possesses serum resistance activity (Nishio et al., 2005) and can promote OM vesicle release in S. Typhimurium (Kitagawa et al., 2010), but none of these roles are confirmed yet for S. Typhi.

Multivalent adhesion molecules (MAM) are outer membrane proteins harbouring 6 or 7 mammalian cell entry domains and are widely found in pathogenic Gram-negative bacteria (Krachler et al., 2011). MAM mediates early interactions with different cell types by providing protein as well as lipid interactions with fibronectin and phosphatidic acid (Krachler et al., 2011). The specificity for certain cell types is thought to be provided by the other adhesins clustered throughout the genome of the bacteria. In S. Typhi, BLASTP analysis revealed that STY1980 harbours about 96% homology with MAM7 of the EPEC strain E. coli O127:H6 (Altschul et al., 1990) and could be implicated during the primary interactions with the intestinal mucosa.

4. Capsule and LPS

S. Typhi produces a group 1 exopolysaccharide known as the Vi antigen. Thus, S. Typhi is one of the few *Salmonella* serovars that get shielded by an extracellular polysaccharide layer constituting the Vi capsule. The Vi polysaccharide is a linear homopolymer of $\alpha(1\rightarrow4)$-2-acetamido-3-O-acetyl-2-deoxy-α-D-galacturonic acid (Heyns et al., 1959) and constitutes the major component of an injectable conjugated vaccine presently used against typhoid fever world-wide (World Health Organization, 2003). Vi has been involved in pathogenicity by evading the host innate immune system as it protects bacteria from phagocytosis and complement-mediated killing (Kossack et al., 1981). The *in vitro* masking of the OAg by the Vi antigen has been known for a long time (Felix & Pitt, 1934 as cited in Robbins & Robbins, 1984), prevents recognition by TLR-4, and limited C3 deposition to the cell surface (Looney & Steigbigel, 1986), which will lead to reduced clearance of the bacteria (Wilson et al., 2011). Vi is preferentially expressed at low osmolarity and early during infection of human macrophages or mice and will be downregulated with the progression of infection (Daigle et al., 2001; Faucher et al., 2006; Janis et al., 2011). The expression of Vi reduces invasion, probably by limiting the access of the T3SS-1 (Arricau et al., 1998; L. Zhao et al., 2001) or by masking other adhesion molecules including Fim. Vi is also important for surviving in macrophages (Hirose et al., 1997). Vi is tightly regulated by its own activator TviA (Hashimoto et al., 1996; Virlogeux et al., 1996), the two-component system OmpR-EnvZ (Pickard et al., 1994), the Rcs system (Arricau et al., 1998; Virlogeux et al., 1996) and repressed by RpoS (Santander et al., 2007).

Lipopolysaccharide (LPS) is the principal component of the outer membrane of Gram-negative bacteria and a major virulence determinant of many pathogens (Raetz & Whitfield,

2002). It is a glycolipid consisting in three structural regions covalently linked: (i) lipid A, also known as endotoxin, a hydrophobic anchor composed of acyl chains linked to phosphorylated N-acetylglucosamine; (ii) the inner and outer core composed of conserved oligosaccharides and; (iii) a variable polysaccharide chain or OAg. *Salmonella* OAg exhibits extensive composition and structural variation and has been divided into 46 O serogroups (Popoff et al., 2001). The O9 antigen of *S.* Typhi is characterized by the presence of a tyvelose residue. In response to acidified macrophage phagosomes, genes activated by the PhoP-PhoQ and PmrA-PmrB systems can modify the global structure of LPS and protect *Salmonella* from being killed by the immune system, notably by antimicrobial peptides (reviewed in Gunn, 2008). Heterogeneity in the length of the OAg repeats has been observed (P. Reeves, 1993) and is important for serum resistance and interaction with host cells (Bravo et al., 2011; Hoare et al., 2006; Hölzer et al., 2009). The *S.* Typhi OAg is essential for serum resistance but is not required for cell invasion (Hoare et al., 2006). Internalization of *S.* Typhi by epithelial cells involves the LPS core (Hoare et al., 2006) which acts as a ligand for CFTR (Lyczak et al., 2001; Pier et al., 1998). The *S.* Typhi LPS core is involved in intracellular replication in macrophages (unpublished data), as observed with *S.* Typhimurium (Nagy et al., 2006; Zenk et al., 2009). *S.* Typhi does not have a bimodal distribution of OAg as it cannot produce very long OAg, consisting of more than 100 repeats of OAg units, because the major regulator Wzz (FepE) is non functional (Raetz & Whitfield, 2002). LPS biosynthesis involves many genes located in different clusters on the chromosome and may be controlled through several regulatory systems (P.R. Reeves et al., 1996). In *S.* Typhi, OAg expression is regulated by RfaH under the control of sigma factor RpoN (Bittner et al., 2002).

5. Secretion systems

5.1 Type 3 secretion systems

S. enterica harbours two distinct type 3 secretion systems (T3SSs) located on SPI-1 (T3SS-1) and SPI-2 (T3SS-2) that are crucial to its virulence along with a flagellar apparatus. T3SSs are complex molecular machines built from more than 20 different proteins, forming a structure similar to a molecular syringe (Kubori et al., 1998, Kimbrough & Miller, 2000 as cited in Sanowar et al., 2010). IM and OM rings are connected by a channel called the needle complex. These structures can inject many protein effectors directly from the bacterial cytoplasm to the cytoplasm of the eukaryotic cells, allowing a direct manipulation of host cellular pathways. The injection process is energized by specific cytoplasmic ATPase and direct contact with the eukaryotic cells is needed in order to activate secretion. Although T3SS are surface-exposed molecules, the lack of specific antibodies against the T3SS in the sera of convalescent patients of typhoid fever (Charles et al., 2010; Harris et al., 2006; Hu et al., 2009) might be a consequence of their tight regulation.

5.1.1 T3SS-1

In order to cause a systemic infection, *Salmonella* must first cross the intestinal epithelial barrier. Conditions found in the intestine, such as low oxygen tension and high osmolarity, are known to induce T3SS-1 of *Salmonella* by the HilA central regulator (Bajaj et al., 1996, Galán & Curtiss, 1990 and Jones & Falkow, 1994 as cited in Altier, 2005). Injection of effectors secreted by the T3SS-1 mediates the invasion of non-phagocytic epithelial cells by *Salmonella* (Galán & Curtiss, 1989; Galán, 1999). Effectors interact with the actin cytoskeleton and induce membrane ruffles around the bacteria allowing its internalisation into epithelial cells. In *S.* Typhi, the contribution of the T3SS-1 during invasion of epithelial cells was confirmed with *invA*, *sipEBCDA* or *iagAB* (*hilAB*) mutants (Galán & Curtiss, 1991; Hermant et al., 1995; Miras et al., 1995). The T3SS-1 of *S.* Typhi may also play a role during the systemic phase of the infection (Haraga et al., 2008; Libby et al., 2010).

5.1.2 T3SS-2

After reaching the epithelial submucosa, *Salmonella* encounters and enters immune system cells like macrophages, dendritic cells and neutrophils. The intracellular environment of these cells promotes induction of the T3SS-2, which is regulated by the SsrA-SsrB two-component regulatory system. Inside cells, bacteria are found in a SCV (*Salmonella*-containing vacuole) and inject T3SS-2 effectors to modify the SCV, alter host pathways and promote intracellular survival (Brumell et al., 2001; Waterman & Holden, 2003; Yu et al., 2004). Although *S.* Typhimurium absolutely requires the T3SS-2 for its intramacrophage survival (Cirillo et al., 1998; Hensel et al., 1998), a complete deletion of this system does not impair survival of *S.* Typhi in human macrophages (Forest et al., 2010). Nevertheless, *S.* Typhi T3SS-2 might be required for survival in other immune cells, as a mutant harbouring a transposon insertion in *ssrB* is disadvantaged in a humanized mouse model (Libby et al., 2010).

5.1.3 Flagella

The flagellar apparatus constitutes a third T3SS that is under the control of a highly organized transcriptional hierarchy involving three promoter classes with *flhDC* being the first activator (Kutsukake et al., 1990 and Karlinsey et al., 2000 as cited in Chevance & Hughes, 2008). In *Salmonella*, each cell harbours 6-8 peritrichous flagella built from more than 25 different proteins (Harshey, 2011). The final structure is composed of a basal body, including a stationary and a moving rotor, an external hook and the filament comprised of flagellin (Harshey, 2011). Secretion of flagellin subunits and motility processes are powered by the proton motive force (Minamino & Namba, 2008 and Paul et al., 2008 as cited in Chevance & Hughes, 2008). Subspecies I, II, IIIa and IV of *Salmonella enterica* are considered biphasic since they can alternatively express FliC or FljB major flagellar subunits in a mechanism known as phase variation (Lederberg & Iino, 1956; Simon et al., 1980). Most *S.* Typhi strains do not possess the *fljB* locus and are monophasic, but some isolates contain a 27 kb linear plasmid harbouring the *fljB*:z66 encoding for a novel flagellin (S. Baker et al., 2007; Frankel et al., 1989). Flagella normally contribute to the virulence through motility and chemotaxis (Macnab, 1999), but can also be implicated during biofilm formation (Crawford et al., 2010a). Flagellin can be detected by TLR-5 present at the cell surface of monocytes, dendritic cells and epithelial cells inducing proinflammatory and adaptive immune responses. In *S.* Typhi, TviA directly downregulates flagellar expression thereby avoiding its early recognition by the intestinal mucosa (Winter et al., 2008). Flagellar genes are involved in survival within macrophages or during the systemic phase of infection (Bäumler et al., 1994; Chan et al., 2005; Klumpp & Fuchs, 2007; Libby et al., 2010; Y. Zhao et al., 2002). Nevertheless, the real contribution of the flagellar apparatus is hard to evaluate since expression of the T3SS-1 is co-regulated with the flagella (Eichelberg & Galán, 2000; Saini et al., 2010). Interestingly, patients harbouring antibodies directed against flagella had uncomplicated typhoid fever, while prevalence of anti-OMP (outer membrane proteins) antibodies was associated with increased ileal perforation rates (Nambiar et al., 2009).

5.2 Type 6 secretion systems

Type 6 secretion systems are newly-discovered structures present in about 25% of sequenced Gram-negative bacterial genomes (Boyer et al., 2009). In *S. enterica* subsp. I, T6SS can be identified within SPI-6 (*S*. Typhi), SPI-19, SPI-20 or SPI-21 (Blondel et al., 2009). T6SS are contractile injection machinery harbouring strong similarities to the tail sheath and spike of bacteriophages (Bönemann et al., 2010). These tubular structures can penetrate eukaryotic as well as prokaryotic membranes in a cell-contact dependant way in order to inject protein effectors. T6SS are often required within phagocytic cells (Ma et al., 2009; Pukatzki et al., 2009), but they can also be implicated in biofilm formation (Aschtgen et al., 2008; Enos-Berlage et al., 2005), colonization of the gastrointestinal tract (Blondel et al., 2010), quorum sensing (Weber et al., 2009) as well as in the delivery of toxins to other cells (Hood et al., 2010). Although *S*. Typhi harbours a pseudogene in a key component of its T6SS, the system is functional and its presence corresponds to an enhanced cytotoxicity toward epithelial cells (M. Wang et al., 2011). T6SS expression is regulated by RcsB, PmrA and Hfq (M. Wang et al., 2011). Its contribution during the interaction with hematopoietic cells should be further studied since a transposon insertion in two genes encoded within SPI-6 showed a competitive disadvantage in humanized mice (Libby et al., 2010).

Surface structure	Role in virulence	Observed for *S*. Typhi	Observed in other serovars
Bcf and Stb	Seroconversion	Harris 2006; Hu 2009	
	Intestinal persistence in mice		Weening 2005
Fim	Binds to mannose, adhesion and invasion of epithelial cells	Satta 1993	Althouse 2003
	Biofilm		Boddicker 2002
Stg	Adhesion to epithelial cells	Forest 2007	
Sth	Long-term infection of mice		Lawley 2006
Saf	Intestinal colonization of swine		Carnell 2007
	Long-term infection of mice		Lawley 2006
Sef	Interaction with macrophages		Edwards 2000
Stc	Intestinal persistence in mice		Weening 2005
Std	Binds to α(1,2)fucose		Chessa 2009
	Intestinal persistence in mice		Weening 2005
Tcf and YaiU	Seroconversion	Harris 2006	
Csg	Seroconversion	Harris 2006	
	Biofilm		Ledeboer 2006
SiiE	Adhesion to apical side of epithelial cells		Gerlach 2008
	Colonization of the gastrointestinal tract		Blondel 2010
BapA	Interaction with intestinal mucosa, bacterial autoaggregation and biofilm		Latasa 2005
ShdA	Seroconversion	Harris 2006	
	Binds to fibronectin		Kingsley 2002
MisL	Intestinal colonization, binds to fibronectin		Dorsey 2005

Surface structure	Role in virulence	Observed for S. Typhi	Observed in other serovars
Type IVB pili	Binds to CFTR, cellular invasion	Pan 2005; Pier 1998	
	Bacterial self-association	Morris 2003b	Morris 2003a
STY0351	Seroconversion	Charles 2010	
	Cell adhesion and binds to laminin	Ghosh 2011	
PagC	Seroconversion	Charles 2010;	
	Survival within macrophages	Harris 2006	Miller 1989
	Serum resistance		Nishio 2005
	OM vesicle release		Kitagawa 2010
Vi Capsule	Host immune system evasion	Kossack 1981; Looney, 1986; Wilson 2011	
	Intramacrophage survival	Hirose 1997	
LPS	Binds to CFTR	Lyczak 2001	
	Antimicrobial peptides resistance	Baker 1999	Gunn 2008
	Serum resistance	Hoare 2006	Bravo 2008
	Intramacrophage survival	Unpublished data	Nagy 2006
T3SS-1	Effectors secretion and invasion of eukaryotic cells	Galán 1991; Hermant 1995	Galán 1989, 1999
T3SS-2	Effectors secretion and intramacrophage survival		Cirillo 1998; Hensel 1998
Flagella	Motility and chemotaxis	Liu 1988	Macnab 1999
	Intramacrophage survival	Unpublished data	Bäumler 1994
	Biofilm formation		Crawford 2010a
T6SS	Colonization of the gastrointestinal tract		Blondel 2010

Table 1. S. Typhi surface structures considered in this review and their roles in virulence.

6. Future perspectives

The multidrug-resistance observed for S. Typhi strains is of great concern since the total number of cases has increased during the last decade (Crump et al., 2004; Pang et al., 1998). There are two crucial lines of defence that should be improved in order to win the combat against typhoid fever : prevention and treatment. The best vaccine would be safe, given in a single dose, offering an efficient and long lasting immunity and remain stable at room temperature. Next generation vaccines have been recently tested in human trials (reviewed in Lindow et al., 2011). The expression of surface structures is tightly coordinated to avoid recognition by the immune system. Nevertheless, we have some clues regarding the structures recognized during typhoid fever (Charles et al., 2010; Harris et al., 2006; Hu et al., 2009). Since antibodies promote killing of S. Typhi (Lindow et al., 2011), a good approach to improve the efficiency of vaccines might be to create an avirulent strain expressing its immunogenic structures on inducible promoters inside antigen presenting cells (S. Wang et al., 2011).

Another strategy in the fight against S. Typhi should be the identification and treatment of the 1-5% infected individuals who become asymptomatic carriers (Crawford et al., 2010b; Parry et al., 2002). This task is complicated as antibiotherapy is often unsuccessful to remove biofilms found in the gallbladder, especially on gallstones, and surgical removal of the gallbladder is usually required but expensive (Crawford et al., 2010b; Prouty et al., 2002) Hence, efforts should be taken to understand the specific structures required for biofilm formation by S. Typhi in order to develop therapies to eliminate typhoid carriage.

Novel strategies are being developed to target surface structures implicated in bacterial pathogenesis as potential treatments (reviewed in Lynch & Wiener-Kronish, 2008). For example, pilicides are small compounds preventing interactions between the OM usher and chaperone-subunits complexes of type 1 pili, hence interfering with fimbrial biogenesis (Pinkner et al., 2006). Since most surface structures of *Salmonella* are expressed by the CUP, targeting the fimbrial ushers might be a useful method to eliminate colonisation and avoid the resulting antimicrobial resistance. Moreover, curlicides are able to interfere with CsgA polymerization as well as type 1 fimbrial biogenesis resulting in the blocking of biofilm accumulation (Cegelski et al., 2009). Similarly, small-molecule inhibitors and inactivating antibodies can target binding or translocation of effectors by T3SS (Hudson et al., 2007; Neely et al., 2005; Nordfelth et al., 2005; Swietnicki et al., 2011). Targeting the capsule or LPS biosynthetic pathways might be a good approach to fight against S. Typhi since there is no corresponding enzyme in its human host (Cipolla et al., 2010; Goller & Seed, 2010).

Finally, understanding the role and function of S. Typhi surface proteins is primordial as these molecules are the first ones to directly interact with host components or cells, leading to a possibility for the development of new strategies to fight typhoid (see Table 1).

7. Acknowledgments

This work was supported by the Canadian Natural Sciences and Engineering Research Council (NSERC) grant number 251114-06. C.G.F. was supported by scholarships from NSERC. We are grateful to Dr. George Szatmari for editing of the manuscript.

8. References

Althouse, C., Patterson, S., Fedorka-Cray, P. & Isaacson, R. E. (2003). Type 1 fimbriae of *Salmonella enterica* serovar Typhimurium bind to enterocytes and contribute to colonization of swine *in vivo*. *Infection and Immunity*, Vol. 71, No. 11, pp. 6446-6452, ISSN 0019-9567

Altier, C. (2005). Genetic and environmental control of *Salmonella* invasion. *Journal of Microbiology*, Vol. 43 Spec No, pp. 85-92, ISSN 1225-8873

Altschul, S. F., Gish, W., Miller, W., Myers, E. W. & Lipman, D. J. (1990). Basic local alignment search tool. *Journal of Molecular Biology*, Vol. 215, No. 3, pp. 403-410, ISSN 0022-2836

Arricau, N., Hermant, D., Waxin, H., Ecobichon, C., Duffey, P. S. & Popoff, M. Y. (1998). The RcsB-RcsC regulatory system of *Salmonella typhi* differentially modulates the expression of invasion proteins, flagellin and Vi antigen in response to osmolarity. *Molecular Microbiology*, Vol. 29, No. 3, pp. 835-850, ISSN 0950-382X

Aschtgen, M. S., Bernard, C. S., De Bentzmann, S., Lloubès, R. & Cascales, E. (2008). SciN is an outer membrane lipoprotein required for type VI secretion in enteroaggregative *Escherichia coli*. *Journal of Bacteriology*, Vol. 190, No. 22, pp. 7523-7531, ISSN 1098-5530

Bajaj, V., Lucas, R. L., Hwang, C. & Lee, C. A. (1996). Co-ordinate regulation of *Salmonella typhimurium* invasion genes by environmental and regulatory factors is mediated by control of *hilA* expression. *Molecular Microbiology*, Vol. 22, No. 4, pp. 703-714, ISSN 0950-382X

Baker, S., Hardy, J., Sanderson, K. E., Quail, M., Goodhead, I., Kingsley, R. A., Parkhill, J., Stocker, B. & Dougan, G. (2007). A novel linear plasmid mediates flagellar variation in *Salmonella* Typhi. *PLoS Pathogens*, Vol. 3, No. 5, p. e59, ISSN 1553-7374

Baker, S. J., Gunn, J. S. & Morona, R. (1999). The *Salmonella typhi* melittin resistance gene *pqaB* affects intracellular growth in PMA-differentiated U937 cells, polymyxin B resistance and lipopolysaccharide. *Microbiology*, Vol. 145, No. 2, pp. 367-378, ISSN 1350-0872

Bäumler, A. J., Kusters, J. G., Stojiljkovic, I. & Heffron, F. (1994). *Salmonella typhimurium* loci involved in survival within macrophages. *Infection and Immunity*, Vol. 62, No. 5, pp. 1623-1630, ISSN 0019-9567

Bishop, A., House, D., Perkins, T., Baker, S., Kingsley, R. A. & Dougan, G. (2008). Interaction of *Salmonella enterica* serovar Typhi with cultured epithelial cells: roles of surface structures in adhesion and invasion. *Microbiology*, Vol. 154, No. 7, pp. 1914-1926, ISSN 1350-0872

Bittner, M., Saldías, S., Estévez, C., Zaldívar, M., Marolda, C. L., Valvano, M. A. & Contreras, I. (2002). O-antigen expression in *Salmonella enterica* serovar Typhi is regulated by nitrogen availability through RpoN-mediated transcriptional control of the *rfaH* gene. *Microbiology*, Vol. 148, No. 12, pp. 3789-3799, ISSN 1350-0872

Blondel, C. J., Jiménez, J. C., Contreras, I. & Santiviago, C. A. (2009). Comparative genomic analysis uncovers 3 novel loci encoding type six secretion systems differentially distributed in *Salmonella* serotypes. *BMC Genomics*, Vol. 10, p. 354, ISSN 1471-2164

Blondel, C. J., Yang, H. J., Castro, B., Chiang, S., Toro, C. S., Zaldívar, M., Contreras, I., Andrews-Polymenis, H. L. & Santiviago, C. A. (2010). Contribution of the type VI secretion system encoded in SPI-19 to chicken colonization by *Salmonella enterica* serotypes Gallinarum and Enteritidis. *PLoS One*, Vol. 5, No. 7, p. e11724, ISSN 1932-6203

Boddicker, J. D., Ledeboer, N. A., Jagnow, J., Jones, B. D. & Clegg, S. (2002). Differential binding to and biofilm formation on, HEp-2 cells by *Salmonella enterica* serovar Typhimurium is dependent upon allelic variation in the *fimH* gene of the *fim* gene cluster. *Molecular Microbiology*, Vol. 45, No. 5, pp. 1255-1265, ISSN 0950-382X

Bönemann, G., Pietrosiuk, A. & Mogk, A. (2010). Tubules and donuts: a type VI secretion story. *Molecular Microbiology*, Vol. 76, No. 4, pp. 815-821, ISSN 1365-2958

Boyer, F., Fichant, G., Berthod, J., Vandenbrouck, Y. & Attree, I. (2009). Dissecting the bacterial type VI secretion system by a genome wide *in silico* analysis: what can be learned from available microbial genomic resources? *BMC Genomics*, Vol. 10, p. 104, ISSN 1471-2164

Bravo, D., Silva, C., Carter, J. A., Hoare, A., Alvarez, S. A., Blondel, C. J., Zaldivar, M., Valvano, M. A. & Contreras, I. (2008). Growth-phase regulation of lipopolysaccharide O-antigen chain length influences serum resistance in serovars of *Salmonella*. *Journal of Medical Microbiology*, Vol. 57, No. 8, pp. 938-946, ISSN 0022-2615

Bravo, D., Hoare, A., Silipo, A., Valenzuela, C., Salinas, C., Alvarez, S. A., Molinaro, A., Valvano, M. A. & Contreras, I. (2011). Different sugar residues of the lipopolysaccharide outer core are required for early interactions of *Salmonella enterica* serovars Typhi and Typhimurium with epithelial cells. *Microbial Pathogenesis*, Vol. 50, No. 2, pp. 70-80, ISSN 1096-1208

Brumell, J. H., Rosenberger, C. M., Gotto, G. T., Marcus, S. L. & Finlay, B. B. (2001). SifA permits survival and replication of *Salmonella typhimurium* in murine macrophages. *Cellular Microbiology*, Vol. 3, No. 2, pp. 75-84, ISSN 1462-5814

Carnell, S. C., Bowen, A., Morgan, E., Maskell, D. J., Wallis, T. S. & Stevens, M. P. (2007). Role in virulence and protective efficacy in pigs of *Salmonella enterica* serovar Typhimurium secreted components identified by signature-tagged mutagenesis. *Microbiology*, Vol. 153, No. 6, pp. 1940-1952, ISSN 1350-0872

Cegelski, L., Pinkner, J. S., Hammer, N. D. & other authors. (2009). Small-molecule inhibitors target *Escherichia coli* amyloid biogenesis and biofilm formation. *Nature Chemical Biology*, Vol. 5, No. 12, pp. 913-919, ISSN 1552-4469

Chan, K., Kim, C. C. & Falkow, S. (2005). Microarray-based detection of *Salmonella enterica* serovar Typhimurium transposon mutants that cannot survive in macrophages and mice. *Infection and Immunity*, Vol. 73, No. 9, pp. 5438-5449, ISSN 0019-9567

Charles, R. C., Harris, J. B., Chase, M. R. & other authors. (2009). Comparative proteomic analysis of the PhoP regulon in *Salmonella enterica* serovar Typhi versus Typhimurium. *PLoS One*, Vol. 4, No. 9, p. e6994, ISSN 1932-6203

Charles, R. C., Sheikh, A., Krastins, B. & other authors. (2010). Characterization of anti-*Salmonella enterica* serotype Typhi antibody responses in bacteremic Bangladeshi patients by an immunoaffinity proteomics-based technology. *Clinical and Vaccine Immunology*, Vol. 17, No. 8, pp. 1188-1195, ISSN 1556-679X

Chessa, D., Winter, M. G., Nuccio, S. P., Tükel, C. & Bäumler, A. J. (2008). RosE represses Std fimbrial expression in *Salmonella enterica* serotype Typhimurium. *Molecular Microbiology*, Vol. 68, No. 3, pp. 573-587, ISSN 1365-2958

Chessa, D., Winter, M. G., Jakomin, M. & Bäumler, A. J. (2009). *Salmonella enterica* serotype Typhimurium Std fimbriae bind terminal alpha(1,2)fucose residues in the cecal mucosa. *Molecular Microbiology*, Vol. 71, No. 4, pp. 864-875, ISSN 1365-2958

Chevance, F. F. & Hughes, K. T. (2008). Coordinating assembly of a bacterial macromolecular machine. *Nature Reviews Microbiology*, Vol. 6, No. 6, pp. 455-465, ISSN 1740-1534

China, B. & Goffaux, F. (1999). Secretion of virulence factors by *Escherichia coli*. *Veterinary Research*, Vol. 30, No. 2-3, pp. 181-202, ISSN 0928-4249

Cipolla, L., Gabrielli, L., Bini, D., Russo, L. & Shaikh, N. (2010). Kdo: a critical monosaccharide for bacteria viability. *Natural Product Reports*, Vol. 27, No. 11, pp. 1618-1629, ISSN 1460-4752

Cirillo, D. M., Valdivia, R. H., Monack, D. M. & Falkow, S. (1998). Macrophage-dependent induction of the *Salmonella* pathogenicity island 2 type III secretion system and its role in intracellular survival. *Molecular Microbiology*, Vol. 30, No. 1, pp. 175-188, ISSN 0950-382X

Craig, L. & Li, J. (2008). Type IV pili: paradoxes in form and function. *Current Opinion in Structural Biology*, Vol. 18, No. 2, pp. 267-277, ISSN 0959-440X

Crawford, R. W., Reeve, K. E. & Gunn, J. S. (2010a). Flagellated but not hyperfimbriated *Salmonella enterica* serovar Typhimurium attaches to and forms biofilms on cholesterol-coated surfaces. *Journal of Bacteriology*, Vol. 192, No. 12, pp. 2981-2990, ISSN 1098-5530

Crawford, R. W., Rosales-Reyes, R., Ramírez-Aguilar, M. e. L., Chapa-Azuela, O., Alpuche-Aranda, C. & Gunn, J. S. (2010b). Gallstones play a significant role in *Salmonella* spp. gallbladder colonization and carriage. *Proceedings of the National Academy of Sciences of the United States of America*, Vol. 107, No. 9, pp. 4353-4358, ISSN 1091-6490

Crump, J. A., Luby, S. P. & Mintz, E. D. (2004). The global burden of typhoid fever. *Bulletin of the World Health Organization*, Vol. 82, No. 5, pp. 346-353, ISSN 0042-9686

Daigle, F., Graham, J. E. & Curtiss III, R. (2001). Identification of *Salmonella typhi* genes expressed within macrophages by selective capture of transcribed sequences (SCOTS). *Molecular Microbiology*, Vol. 41, No. 5, pp. 1211-1222, ISSN 0950-382X

Dorsey, C. W., Laarakker, M. C., Humphries, A. D., Weening, E. H. & Bäumler, A. J. (2005). *Salmonella enterica* serotype Typhimurium MisL is an intestinal colonization factor that binds fibronectin. *Molecular Microbiology*, Vol. 57, No. 1, pp. 196-211, ISSN 0950-382X

Duguid, J. P., Anderson, E. S. & Campbell, I. (1966). Fimbriae and adhesive properties in Salmonellae. *Journal of Pathology and Bacteriology*, Vol. 92, No. 1, pp. 107-138, ISSN 0368-3494

Edwards, R. A., Schifferli, D. M. & Maloy, S. R. (2000). A role for *Salmonella* fimbriae in intraperitoneal infections. *Proceedings of the National Academy of Sciences of the United States of America*, Vol. 97, No. 3, pp. 1258-1262, ISSN 0027-8424

Eichelberg, K. & Galán, J. E. (2000). The flagellar sigma factor FliA (sigma(28)) regulates the expression of *Salmonella* genes associated with the centisome 63 type III secretion system. *Infection and Immunity*, Vol. 68, No. 5, pp. 2735-2743, ISSN 0019-9567

Enos-Berlage, J. L., Guvener, Z. T., Keenan, C. E. & McCarter, L. L. (2005). Genetic determinants of biofilm development of opaque and translucent *Vibrio parahaemolyticus*. *Molecular Microbiology*, Vol. 55, No. 4, pp. 1160-1182, ISSN 0950-382X

Faucher, S. P., Porwollik, S., Dozois, C. M., McClelland, M. & Daigle, F. (2006). Transcriptome of *Salmonella enterica* serovar Typhi within macrophages revealed through the selective capture of transcribed sequences. *Proceedings of the National Academy of Sciences of the United States of America*, Vol. 103, No. 6, pp. 1906-1911, ISSN 0027-8424

Felix, A. & Pitt, R. M. (1934). A new antigen of *B. typhosus*. *The Lancet*, Vol. 224, No. 5787, pp. 186-191, ISSN 0140-6736

Forest, C., Faucher, S. P., Poirier, K., Houle, S., Dozois, C. M. & Daigle, F. (2007). Contribution of the *stg* fimbrial operon of *Salmonella enterica* serovar Typhi during interaction with human cells. *Infection and Immunity*, Vol. 75, No. 11, pp. 5264-5271, ISSN 0019-9567

Forest, C. G., Ferraro, E., Sabbagh, S. C. & Daigle, F. (2010). Intracellular survival of *Salmonella enterica* serovar Typhi in human macrophages is independent of *Salmonella* pathogenicity island (SPI)-2. *Microbiology*, Vol. 156, No. 12, pp. 3689-3698, ISSN 1465-2080

Frankel, G., Newton, S. M., Schoolnik, G. K. & Stocker, B. A. (1989). Intragenic recombination in a flagellin gene: characterization of the H1-j gene of *Salmonella typhi*. *EMBO Journal*, Vol. 8, No. 10, pp. 3149-3152, ISSN 0261-4189

Galán, J. E. & Curtiss III, R. (1989). Cloning and molecular characterization of genes whose products allow *Salmonella typhimurium* to penetrate tissue culture cells. *Proceedings of the National Academy of Sciences of the United States of America*, Vol. 86, No. 16, pp. 6383-6387, ISSN 0027-8424

Galán, J. E. & Curtiss III, R. (1990). Expression of *Salmonella typhimurium* genes required for invasion is regulated by changes in DNA supercoiling. *Infection and Immunity*, Vol. 58, No. 6, pp. 1879-1885, ISSN 0019-9567

Galán, J. E. & Curtiss III, R. (1991). Distribution of the *invA, -B, -C,* and *-D* genes of *Salmonella typhimurium* among other *Salmonella* serovars: *invA* mutants of *Salmonella typhi* are deficient for entry into mammalian cells. *Infection and Immunity*, Vol. 59, No. 9, pp. 2901-2908, ISSN 0019-9567

Galán, J. E. (1999). Interaction of *Salmonella* with host cells through the centisome 63 type III secretion system. *Current Opinion in Microbiology*, Vol. 2, No. 1, pp. 46-50, ISSN 1369-5274

Gerlach, R. G., Jäckel, D., Stecher, B., Wagner, C., Lupas, A., Hardt, W. D. & Hensel, M. (2007). *Salmonella* Pathogenicity Island 4 encodes a giant non-fimbrial adhesin and the cognate type 1 secretion system. *Cellular Microbiology*, Vol. 9, No. 7, pp. 1834-1850, ISSN 1462-5814

Gerlach, R. G., Cláudio, N., Rohde, M., Jäckel, D., Wagner, C. & Hensel, M. (2008). Cooperation of *Salmonella* pathogenicity islands 1 and 4 is required to breach epithelial barriers. *Cellular Microbiology*, Vol. 10, No. 11, pp. 2364-2376, ISSN 1462-5822

Ghosh, S., Chakraborty, K., Nagaraja, T., Basak, S., Koley, H., Dutta, S., Mitra, U. & Das, S. (2011). An adhesion protein of *Salmonella enterica* serovar Typhi is required for pathogenesis and potential target for vaccine development. *Proceedings of the National Academy of Sciences of the United States of America*, Vol. 108, No. 8, pp. 3348-3353, ISSN 1091-6490

Goller, C. C. & Seed, P. C. (2010). High-throughput identification of chemical inhibitors of *E. coli* Group 2 capsule biogenesis as anti-virulence agents. *PLoS One*, Vol. 5, No. 7, p. e11642, ISSN 1932-6203

Grosskinsky, U., Schütz, M., Fritz, M., Schmid, Y., Lamparter, M. C., Szczesny, P., Lupas, A. N., Autenrieth, I. B. & Linke, D. (2007). A conserved glycine residue of trimeric autotransporter domains plays a key role in *Yersinia* adhesin A autotransport. *Journal of Bacteriology*, Vol. 189, No. 24, pp. 9011-9019, ISSN 1098-5530

Gunn, J. S. (2008). The *Salmonella* PmrAB regulon: lipopolysaccharide modifications, antimicrobial peptide resistance and more. *Trends in Microbiology*, Vol. 16, No. 6, pp. 284-290, ISSN 0966-842X

Hammar, M., Bian, Z. & Normark, S. (1996). Nucleator-dependent intercellular assembly of adhesive curli organelles in *Escherichia coli*. *Proceedings of the National Academy of Sciences of the United States of America*, Vol. 93, No. 13, pp. 6562-6566, ISSN 0027-8424

Haraga, A., Ohlson, M. B. & Miller, S. I. (2008). Salmonellae interplay with host cells. *Nature Reviews Microbiology*, Vol. 6, No. 1, pp. 53-66, ISSN 1740-1534

Harris, J. B., Baresch-Bernal, A., Rollins, S. M. & other authors. (2006). Identification of *in vivo*-induced bacterial protein antigens during human infection with *Salmonella enterica* serovar Typhi. *Infection and Immunity*, Vol. 74, No. 9, pp. 5161-5168, ISSN 0019-9567

Harshey, R. M. (2011). New insights into the role and formation of flagella in *Salmonella*, In: *Salmonella: from genome to function*, S. Porwollik (Ed.), pp. 163-186, Caister Academic Press, ISBN 978-1-904455-73-8, Norfolk

Hashimoto, Y., Khan, A. Q. & Ezaki, T. (1996). Positive autoregulation of *vipR* expression in ViaB region-encoded Vi antigen of *Salmonella typhi*. *Journal of Bacteriology*, Vol. 178, No. 5, pp. 1430-1436, ISSN 0021-9193

Hensel, M., Shea, J. E., Waterman, S. R. & other authors. (1998). Genes encoding putative effector proteins of the type III secretion system of *Salmonella* pathogenicity island 2 are required for bacterial virulence and proliferation in macrophages. *Molecular Microbiology*, Vol. 30, No. 1, pp. 163-174, ISSN 0950-382X

Hermant, D., Ménard, R., Arricau, N., Parsot, C. & Popoff, M. Y. (1995). Functional conservation of the *Salmonella* and *Shigella* effectors of entry into epithelial cells. *Molecular Microbiology*, Vol. 17, No. 4, pp. 781-789, ISSN 0950-382X

Heyns, K., Kiessling, G., Lindenberg, W., Paulsen, H. & Webster, M. E. (1959). D-Galaktosaminuornsaure (2-amino-2-desoxy-D-galakturonsaure) als Baustein des Vi-Antigens. *Chemische Berichte*, Vol. 92, No. 10, pp. 2435-2437, ISSN 1099-0682

Hirose, K., Ezaki, T., Miyake, M., Li, T., Khan, A. Q., Kawamura, Y., Yokoyama, H. & Takami, T. (1997). Survival of Vi-capsulated and Vi-deleted *Salmonella typhi* strains in cultured macrophage expressing different levels of CD14 antigen. *FEMS Microbiology Letters*, Vol. 147, No. 2, pp. 259-265, ISSN 0378-1097

Hoare, A., Bittner, M., Carter, J., Alvarez, S., Zaldívar, M., Bravo, D., Valvano, M. A. & Contreras, I. (2006). The outer core lipopolysaccharide of *Salmonella enterica* serovar Typhi is required for bacterial entry into epithelial cells. *Infection and Immunity*, Vol. 74, No. 3, pp. 1555-1564, ISSN 0019-9567

Holt, K. E., Thomson, N. R., Wain, J. & other authors. (2009). Pseudogene accumulation in the evolutionary histories of *Salmonella enterica* serovars Paratyphi A and Typhi. *BMC Genomics*, Vol. 10, p. 36, ISSN 1471-2164

Hölzer, S. U., Schlumberger, M. C., Jäckel, D. & Hensel, M. (2009). Effect of the O-antigen length of lipopolysaccharide on the functions of Type III secretion systems in *Salmonella enterica*. *Infection and Immunity*, Vol. 77, No. 12, pp. 5458-5470, ISSN 1098-5522

Hood, R. D., Singh, P., Hsu, F. & other authors. (2010). A type VI secretion system of *Pseudomonas aeruginosa* targets a toxin to bacteria. *Cell Host Microbe*, Vol. 7, No. 1, pp. 25-37, ISSN 1934-6069

Hu, Y., Cong, Y., Li, S., Rao, X., Wang, G. & Hu, F. (2009). Identification of *in vivo* induced protein antigens of *Salmonella enterica* serovar Typhi during human infection. *Science in China Series C, Life Sciences*, Vol. 52, No. 10, pp. 942-948, ISSN 1862-2798

Hudson, D. L., Layton, A. N., Field, T. R., Bowen, A. J., Wolf-Watz, H., Elofsson, M., Stevens, M. P. & Galyov, E. E. (2007). Inhibition of type III secretion in *Salmonella enterica* serovar Typhimurium by small-molecule inhibitors. *Antimicrobial Agents and Chemotherapy*, Vol. 51, No. 7, pp. 2631-2635, ISSN 0066-4804

Humphries, A. D., Raffatellu, M., Winter, S. & other authors. (2003). The use of flow cytometry to detect expression of subunits encoded by 11 *Salmonella enterica* serotype Typhimurium fimbrial operons. *Molecular Microbiology*, Vol. 48, No. 5, pp. 1357-1376, ISSN 0950-382X

Jacob-Dubuisson, F., Striker, R. & Hultgren, S. J. (1994). Chaperone-assisted self-assembly of pili independent of cellular energy. *Journal of Biological Chemistry*, Vol. 269, No. 17, pp. 12447-12455, ISSN 0021-9258

Janis, C., Grant, A. J., McKinley, T. J., Morgan, F. J., John, V. F., Houghton, J., Kingsley, R. A., Dougan, G. & Mastroeni, P. (2011). *In vivo* regulation of the Vi antigen in *Salmonella* and induction of immune responses with an *in vivo*-inducible promoter. *Infection and Immunity*, Vol. 79, No. 6, pp. 2481-2488, ISSN 1098-5522

Jonas, K., Tomenius, H., Kader, A., Normark, S., Römling, U., Belova, L. M. & Melefors, O. (2007). Roles of curli, cellulose and BapA in *Salmonella* biofilm morphology studied by atomic force microscopy. *BMC Microbiology*, Vol. 7, p. 70, ISSN 1471-2180

Jones, B. D. & Falkow, S. (1994). Identification and characterization of a *Salmonella typhimurium* oxygen-regulated gene required for bacterial internalization. *Infection and Immunity*, Vol. 62, No. 9, pp. 3745-3752, ISSN 0019-9567

Karlinsey, J. E., Tanaka, S., Bettenworth, V., Yamaguchi, S., Boos, W., Aizawa, S. I. & Hughes, K. T. (2000). Completion of the hook-basal body complex of the *Salmonella typhimurium* flagellum is coupled to FlgM secretion and *fliC* transcription. *Molecular Microbiology*, Vol. 37, No. 5, pp. 1220-1231, ISSN 0950-382X

Kidgell, C., Reichard, U., Wain, J., Linz, B., Torpdahl, M., Dougan, G. & Achtman, M. (2002). *Salmonella typhi*, the causative agent of typhoid fever, is approximately 50,000 years old. *Infection, genetic and evolution*, Vol. 2, No. 1, pp. 39-45, ISSN 1567-1348

Kimbrough, T. & Miller, S. (2000). Contribution of *Salmonella typhimurium* type III secretion components to needle complex formation. *Proceedings of the National Academy of Sciences of the United States of America*, Vol. 97, No. 20, pp. 11008-11013, ISSN 0027-8424

Kingsley, R. A., Santos, R. L., Keestra, A. M., Adams, L. G. & Bäumler, A. J. (2002). *Salmonella enterica* serotype Typhimurium ShdA is an outer membrane fibronectin-binding protein that is expressed in the intestine. *Molecular Microbiology*, Vol. 43, No. 4, pp. 895-905, ISSN 0950-382X

Kitagawa, R., Takaya, A., Ohya, M., Mizunoe, Y., Takade, A., Yoshida, S., Isogai, E. & Yamamoto, T. (2010). Biogenesis of *Salmonella enterica* serovar Typhimurium membrane vesicles provoked by induction of PagC. *Journal of Bacteriology*, Vol. 192, No. 21, pp. 5645-5656, ISSN 1098-5530

Klumpp, J. & Fuchs, T. M. (2007). Identification of novel genes in genomic islands that contribute to *Salmonella typhimurium* replication in macrophages. *Microbiology*, Vol. 153, No. 4, pp. 1207-1220, ISSN 1350-0872

Korea, C. G., Ghigo, J. M. & Beloin, C. (2011). The sweet connection: Solving the riddle of multiple sugar-binding fimbrial adhesins in *Escherichia coli*: Multiple *E. coli* fimbriae form a versatile arsenal of sugar-binding lectins potentially involved in surface-colonisation and tissue tropism. *Bioessays*, Vol. 33, No. 4, pp. 300-311, ISSN 1521-1878

Kossack, R. E., Guerrant, R. L., Densen, P., Schadelin, J. & Mandell, G. L. (1981). Diminished neutrophil oxidative metabolism after phagocytosis of virulent *Salmonella typhi*. *Infection and Immunity*, Vol. 31, No. 2, pp. 674-678, ISSN 0019-9567

Krachler, A. M., Ham, H. & Orth, K. (2011). Outer membrane adhesion factor multivalent adhesion molecule 7 initiates host cell binding during infection by Gram-negative pathogens. *Proceedings of the National Academy of Sciences of the United States of America*, Vol. 108, No. 28, pp. 11614-11619, ISSN 1091-6490

Kubori, T., Matsushima, Y., Nakamura, D., Uralil, J., Lara-Tejero, M., Sukhan, A., Galan, J. E. & Aizawa, S. I. (1998). Supramolecular structure of the *Salmonella typhimurium* type III protein secretion system. *Science*, Vol. 280, No. 5363, pp. 602-605, ISSN 0036-8075

Kutsukake, K., Ohya, Y. & Iino, T. (1990). Transcriptional analysis of the flagellar regulon of *Salmonella typhimurium*. *Journal of Bacteriology*, Vol. 172, No. 2, pp. 741-747, ISSN 0021-9193

Latasa, C., Roux, A., Toledo-Arana, A., Ghigo, J. M., Gamazo, C., Penadés, J. R. & Lasa, I. (2005). BapA, a large secreted protein required for biofilm formation and host colonization of *Salmonella enterica* serovar Enteritidis. *Molecular Microbiology*, Vol. 58, No. 5, pp. 1322-1339, ISSN 0950-382X

Lawley, T. D., Chan, K., Thompson, L. J., Kim, C. C., Govoni, G. R. & Monack, D. M. (2006). Genome-wide screen for *Salmonella* genes required for long-term systemic infection of the mouse. *PLoS Pathogens*, Vol. 2, No. 2, p. e11, ISSN 1553-7374

Ledeboer, N. A., Frye, J. G., McClelland, M. & Jones, B. D. (2006). *Salmonella enterica* serovar Typhimurium requires the Lpf, Pef, and Tafi fimbriae for biofilm formation on HEp-2 tissue culture cells and chicken intestinal epithelium. *Infection and Immunity*, Vol. 74, No. 6, pp. 3156-3169, ISSN 0019-9567

Lederberg, J. & Iino, T. (1956). Phase Variation in *Salmonella*. *Genetics*, Vol. 41, No. 5, pp. 743-757, ISSN 0016-6731

Libby, S. J., Brehm, M. A., Greiner, D. L. & other authors. (2010). Humanized nonobese diabetic-scid IL2rgammanull mice are susceptible to lethal *Salmonella* Typhi infection. *Proceedings of the National Academy of Sciences of the United States of America*, Vol. 107, No. 35, pp. 15589-15594, ISSN 1091-6490

Lindow, J. C., Fimlaid, K. A., Bunn, J. Y. & Kirkpatrick, B. D. (2011). Antibodies in Action: Role of Human Opsonins in Killing *Salmonella enterica* Serovar Typhi. *Infection and Immunity*, Vol. 79, No. 8, pp. 3188-3194, ISSN 1098-5522

Liu, S. L., Ezaki, T., Miura, H., Matsui, K. & Yabuuchi, E. (1988). Intact motility as a *Salmonella typhi* invasion-related factor. *Infection and Immunity*, Vol. 56, No. 8, pp. 1967-1973, ISSN 0019-9567

Looney, R. J. & Steigbigel, R. T. (1986). Role of the Vi antigen of *Salmonella typhi* in resistance to host defense *in vitro*. *Journal of Laboratory and Clinical Medicine*, Vol. 108, No. 5, pp. 506-516, ISSN 0022-2143

Lyczak, J. B., Zaidi, T. S., Grout, M., Bittner, M., Contreras, I. & Pier, G. B. (2001). Epithelial cell contact-induced alterations in *Salmonella enterica* serovar Typhi lipopolysaccharide are critical for bacterial internalization. *Cellular Microbiology*, Vol. 3, No. 11, pp. 763-772, ISSN 1462-5814

Lyczak, J. B. & Pier, G. B. (2002). *Salmonella enterica* serovar Typhi modulates cell surface expression of its receptor, the cystic fibrosis transmembrane conductance regulator, on the intestinal epithelium. *Infection and Immunity*, Vol. 70, No. 11, pp. 6416-6423, ISSN 0019-9567

Lynch, S. V. & Wiener-Kronish, J. P. (2008). Novel strategies to combat bacterial virulence. *Current Opinion in Critical Care*, Vol. 14, No. 5, pp. 593-599, ISSN 1531-7072

Ma, A. T., McAuley, S., Pukatzki, S. & Mekalanos, J. J. (2009). Translocation of a *Vibrio cholerae* type VI secretion effector requires bacterial endocytosis by host cells. *Cell Host Microbe*, Vol. 5, No. 3, pp. 234-243, ISSN 1934-6069

Macnab, R. M. (1999). The bacterial flagellum: reversible rotary propellor and type III export apparatus. *Journal of Bacteriology*, Vol. 181, No. 23, pp. 7149-7153, ISSN 0021-9193

Main-Hester, K. L., Colpitts, K. M., Thomas, G. A., Fang, F. C. & Libby, S. J. (2008). Coordinate regulation of *Salmonella* pathogenicity island 1 (SPI1) and SPI4 in

Salmonella enterica serovar Typhimurium. *Infection and Immunity*, Vol. 76, No. 3, pp. 1024-1035, ISSN 1098-5522

Mattick, J. S. (2002). Type IV pili and twitching motility. *Annual Review of Microbiology*, Vol. 56, pp. 289-314, ISSN 0066-4227

Miller, S. I., Kukral, A. M. & Mekalanos, J. J. (1989). A two-component regulatory system (*phoP phoQ*) controls *Salmonella typhimurium* virulence. *Proceedings of the National Academy of Sciences of the United States of America*, Vol. 86, No. 13, pp. 5054-5058, ISSN 0027-8424

Minamino, T. & Namba, K. (2008). Distinct roles of the FliI ATPase and proton motive force in bacterial flagellar protein export. *Nature*, Vol. 451, No. 7177, pp. 485-488, ISSN 1476-4687

Miras, I., Hermant, D., Arricau, N. & Popoff, M. Y. (1995). Nucleotide sequence of *iagA* and *iagB* genes involved in invasion of HeLa cells by *Salmonella enterica* subsp. *enterica* ser. Typhi. *Research in Microbiology*, Vol. 146, No. 1, pp. 17-20, ISSN 0923-2508

Miyake, M., Zhao, L., Ezaki, T. & other authors. (1998). Vi-deficient and nonfimbriated mutants of *Salmonella typhi* agglutinate human blood type antigens and are hyperinvasive. *FEMS Microbiology Letters*, Vol. 161, No. 1, pp. 75-82, ISSN 0378-1097

Morgan, E., Campbell, J. D., Rowe, S. C., Bispham, J., Stevens, M. P., Bowen, A. J., Barrow, P. A., Maskell, D. J. & Wallis, T. S. (2004). Identification of host-specific colonization factors of *Salmonella enterica* serovar Typhimurium. *Molecular Microbiology*, Vol. 54, No. 4, pp. 994-1010, ISSN 0950-382X

Morris, C., Tam, C. K., Wallis, T. S., Jones, P. W. & Hackett, J. (2003a). *Salmonella enterica* serovar Dublin strains which are Vi antigen-positive use type IVB pili for bacterial self-association and human intestinal cell entry. *Microbial Pathogenesis*, Vol. 35, No. 6, pp. 279-284, ISSN 0882-4010

Morris, C., Yip, C. M., Tsui, I. S., Wong, D. K. & Hackett, J. (2003b). The shufflon of *Salmonella enterica* serovar Typhi regulates type IVB pilus-mediated bacterial self-association. *Infection and Immunity*, Vol. 71, No. 3, pp. 1141-1146, ISSN 0019-9567

Nagy, G., Danino, V., Dobrindt, U., Pallen, M., Chaudhuri, R., Emödy, L., Hinton, J. C. & Hacker, J. (2006). Down-regulation of key virulence factors makes the *Salmonella enterica* serovar Typhimurium *rfaH* mutant a promising live-attenuated vaccine candidate. *Infection and Immunity*, Vol. 74, No. 10, pp. 5914-5925, ISSN 0019-9567

Nambiar, M., Harish, B. N., Mangilal, V., Pai, D. & Parija, S. C. (2009). Immunoblot analysis of sera in uncomplicated typhoid fever & with typhoid ileal perforation. *Indian Journal of Medical Research*, Vol. 129, No. 4, pp. 432-437, ISSN 0971-5916

Neely, A. N., Holder, I. A., Wiener-Kronish, J. P. & Sawa, T. (2005). Passive anti-PcrV treatment protects burned mice against *Pseudomonas aeruginosa* challenge. *Burns*, Vol. 31, No. 2, pp. 153-158, ISSN 0305-4179

Nishimura, K., Tajima, N., Yoon, Y. H., Park, S. Y. & Tame, J. R. (2010). Autotransporter passenger proteins: virulence factors with common structural themes. *Journal of Molecular Medicine*, Vol. 88, No. 5, pp. 451-458, ISSN 1432-1440

Nishio, M., Okada, N., Miki, T., Haneda, T. & Danbara, H. (2005). Identification of the outer-membrane protein PagC required for the serum resistance phenotype in *Salmonella enterica* serovar Choleraesuis. *Microbiology*, Vol. 151, No. 3, pp. 863-873, ISSN 1350-0872

Nishiyama, M., Ishikawa, T., Rechsteiner, H. & Glockshuber, R. (2008). Reconstitution of pilus assembly reveals a bacterial outer membrane catalyst. *Science*, Vol. 320, No. 5874, pp. 376-379, ISSN 1095-9203

Nordfelth, R., Kauppi, A. M., Norberg, H. A., Wolf-Watz, H. & Elofsson, M. (2005). Small-molecule inhibitors specifically targeting type III secretion. *Infection and Immunity*, Vol. 73, No. 5, pp. 3104-3114, ISSN 0019-9567

Norris, T. & Bäumler, A. (1999). Phase variation of the *lpf* operon is a mechanism to evade cross-immunity between *Salmonella* serotypes. *Proceedings of the National Academy of Sciences of the United States of America*, Vol. 96, No. 23, pp. 13393-13398, ISSN 0027-8424

Nuccio, S. P. & Bäumler, A. J. (2007). Evolution of the chaperone/usher assembly pathway: fimbrial classification goes Greek. *Microbiology and Molecular Biology Reviews*, Vol. 71, No. 4, pp. 551-575, ISSN 1092-2172

Nuccio, S. P., Thomson, N. R., Fookes, M. C. & Baümler, A. J. (2011). Fimbrial signature arrangements in *Salmonella* In: *Salmonella: from genome to function*, S. Porwollick (Ed.), pp. 149-161, Caister Academic Press, ISBN 978-1-904455-73-8, Norfolk

Pan, Q., Zhang, X. L., Wu, H. Y., He, P. W., Wang, F., Zhang, M. S., Hu, J. M., Xia, B. & Wu, J. (2005). Aptamers that preferentially bind type IVB pili and inhibit human monocytic-cell invasion by *Salmonella enterica* serovar Typhi. *Antimicrobial Agents and Chemotherapy*, Vol. 49, No. 10, pp. 4052-4060, ISSN 0066-4804

Pang, T., Levine, M. M., Ivanoff, B., Wain, J. & Finlay, B. B. (1998). Typhoid fever--important issues still remain. *Trends in Microbiology*, Vol. 6, No. 4, pp. 131-133, ISSN 0966-842X

Parkhill, J., Dougan, G., James, K. D. & other authors. (2001). Complete genome sequence of a multiple drug resistant *Salmonella enterica* serovar Typhi CT18. *Nature*, Vol. 413, No. 6858, pp. 848-852, ISSN 0028-0836

Parry, C. M., Hien, T. T., Dougan, G., White, N. J. & Farrar, J. J. (2002). Typhoid fever. *New England Journal of Medicine*, Vol. 347, No. 22, pp. 1770-1782, ISSN 1533-4406

Paul, K., Erhardt, M., Hirano, T., Blair, D. F. & Hughes, K. T. (2008). Energy source of flagellar type III secretion. *Nature*, Vol. 451, No. 7177, pp. 489-492, ISSN 1476-4687

Pickard, D., Li, J., Roberts, M., Maskell, D., Hone, D., Levine, M., Dougan, G. & Chatfield, S. (1994). Characterization of defined *ompR* mutants of *Salmonella typhi*: *ompR* is involved in the regulation of Vi polysaccharide expression. *Infection and Immunity*, Vol. 62, No. 9, pp. 3984-3993, ISSN 0019-9567

Pier, G. B., Grout, M., Zaidi, T., Meluleni, G., Mueschenborn, S. S., Banting, G., Ratcliff, R., Evans, M. J. & Colledge, W. H. (1998). *Salmonella typhi* uses CFTR to enter intestinal epithelial cells. *Nature*, Vol. 393, No. 6680, pp. 79-82, ISSN 0028-0836

Pinkner, J. S., Remaut, H., Buelens, F. & other authors. (2006). Rationally designed small compounds inhibit pilus biogenesis in uropathogenic bacteria. *Proceedings of the National Academy of Sciences of the United States of America*, Vol. 103, No. 47, pp. 17897-17902, ISSN 0027-8424

Popoff, M. Y., Bockemühl, J., Brenner, F. W. & Gheesling, L. L. (2001). Supplement 2000 (no. 44) to the Kauffmann-White scheme. *Research in Microbiology*, Vol. 152, No. 10, pp. 907-909, ISSN 0923-2508

Prouty, A. M., Schwesinger, W. H. & Gunn, J. S. (2002). Biofilm formation and interaction with the surfaces of gallstones by *Salmonella* spp. *Infection and Immunity*, Vol. 70, No. 5, pp. 2640-2649, ISSN 0019-9567

Pukatzki, S., McAuley, S. B. & Miyata, S. T. (2009). The type VI secretion system: translocation of effectors and effector-domains. *Current Opinion in Microbiology*, Vol. 12, No. 1, pp. 11-17, ISSN 1879-0364

Raetz, C. R. & Whitfield, C. (2002). Lipopolysaccharide endotoxins. *Annual Review of Biochemistry*, Vol. 71, pp. 635-700, ISSN 0066-4154

Reeves, P. (1993). Evolution of *Salmonella* O antigen variation by interspecific gene transfer on a large scale. *Trends in Genetics*, Vol. 9, No. 1, pp. 17-22, ISSN 0168-9525

Reeves, P. R., Hobbs, M., Valvano, M. A. & other authors. (1996). Bacterial polysaccharide synthesis and gene nomenclature. *Trends in Microbiology*, Vol. 4, No. 12, pp. 495-503, ISSN 0966-842X

Robbins, J. D. & Robbins, J. B. (1984). Reexamination of the protective role of the capsular polysaccharide (Vi antigen) of *Salmonella typhi*. *Journal of Infectious Diseases*, Vol. 150, No. 3, pp. 436-449, ISSN 0022-1899

Römling, U., Bokranz, W., Rabsch, W., Zogaj, X., Nimtz, M. & Tschäpe, H. (2003). Occurrence and regulation of the multicellular morphotype in *Salmonella* serovars important in human disease. *International Journal of Medical Microbiology*, Vol. 293, No. 4, pp. 273-285, ISSN 1438-4221

Sabbagh, S. C., Forest, C. G., Lepage, C., Leclerc, J. M. & Daigle, F. (2010). So similar, yet so different: uncovering distinctive features in the genomes of *Salmonella enterica* serovars Typhimurium and Typhi. *FEMS Microbiology Letters*, Vol. 305, No. 1, pp. 1-13, ISSN 1574-6968

Saini, S., Slauch, J. M., Aldridge, P. D. & Rao, C. V. (2010). Role of cross talk in regulating the dynamic expression of the flagellar *Salmonella* pathogenicity island 1 and type 1 fimbrial genes. *Journal of Bacteriology*, Vol. 192, No. 21, pp. 5767-5777, ISSN 1098-5530

Salih, O., Remaut, H., Waksman, G. & Orlova, E. V. (2008). Structural analysis of the Saf pilus by electron microscopy and image processing. *Journal of Molecular Biology*, Vol. 379, No. 1, pp. 174-187, ISSN 1089-8638

Sanowar, S., Singh, P., Pfuetzner, R. A. & other authors. (2010). Interactions of the transmembrane polymeric rings of the *Salmonella enterica* serovar Typhimurium type III secretion system. *MBio*, Vol. 1, No. 3, ISSN 2150-7511

Santander, J., Wanda, S. Y., Nickerson, C. A. & Curtiss III, R. (2007). Role of RpoS in fine-tuning the synthesis of Vi capsular polysaccharide in *Salmonella enterica* serotype Typhi. *Infection and Immunity*, Vol. 75, No. 3, pp. 1382-1392, ISSN 0019-9567

Satta, G., Ingianni, L., Muscas, P., Rossolini, A. & Pompei, R. (1993). The pathogenicity determinants of *Salmonella typhi*: potential role of fimbrial structures, In: *Biology of Salmonella*, F. Cabello, C. Hormaeche, P. Mastroeni & L. Bonina (Eds.), pp. 83-90, Plenum Press, ISBN 0306444925, New York

Sauer, F. G., Pinkner, J. S., Waksman, G. & Hultgren, S. J. (2002). Chaperone priming of pilus subunits facilitates a topological transition that drives fiber formation. *Cell*, Vol. 111, No. 4, pp. 543-551, ISSN 0092-8674

Simon, M., Zieg, J., Silverman, M., Mandel, G. & Doolittle, R. (1980). Phase variation: evolution of a controlling element. *Science*, Vol. 209, No. 4463, pp. 1370-1374, ISSN 0036-8075

Swietnicki, W., Carmany, D., Retford, M., Guelta, M., Dorsey, R., Bozue, J., Lee, M. S. & Olson, M. A. (2011). Identification of small-molecule inhibitors of *Yersinia pestis* Type III secretion system YscN ATPase. *PLoS One*, Vol. 6, No. 5, p. e19716, ISSN 1932-6203

Thankavel, K., Shah, A. H., Cohen, M. S., Ikeda, T., Lorenz, R. G., Curtiss III, R. & Abraham, S. N. (1999). Molecular basis for the enterocyte tropism exhibited by *Salmonella typhimurium* type 1 fimbriae. *Journal of Biological Chemistry*, Vol. 274, No. 9, pp. 5797-5809, ISSN 0021-9258

Townsend, S. M., Kramer, N. E., Edwards, R. & other authors. (2001). *Salmonella enterica* serovar Typhi possesses a unique repertoire of fimbrial gene sequences. *Infection and Immunity*, Vol. 69, No. 5, pp. 2894-2901, ISSN 0019-9567

Tsui, I. S., Yip, C. M., Hackett, J. & Morris, C. (2003). The type IVB pili of *Salmonella enterica* serovar Typhi bind to the cystic fibrosis transmembrane conductance regulator. *Infection and Immunity*, Vol. 71, No. 10, pp. 6049-6050, ISSN 0019-9567

Virlogeux, I., Waxin, H., Ecobichon, C., Lee, J. O. & Popoff, M. Y. (1996). Characterization of the *rcsA* and *rcsB* genes from *Salmonella typhi*: *rcsB* through *tviA* is involved in regulation of Vi antigen synthesis. *Journal of Bacteriology*, Vol. 178, No. 6, pp. 1691-1698, ISSN 0021-9193

Wang, F., Zhang, X. L., Zhou, Y., Ye, L., Qi, Z. & Wu, J. (2005). Type IVB piliated *Salmonella typhi* enhance IL-6 and NF-kappaB production in human monocytic THP-1 cells through activation of protein kinase C. *Immunobiology*, Vol. 210, No. 5, pp. 283-293, ISSN 0171-2985

Wang, M., Luo, Z., Du, H., Xu, S., Ni, B., Zhang, H., Sheng, X., Xu, H. & Huang, X. (2011). Molecular Characterization of a Functional Type VI Secretion System in *Salmonella enterica* serovar Typhi. *Current Microbiology*, Vol. 63, No. 1, pp. 22-31, ISSN 1432-0991

Wang, S., Li, Y., Shi, H., Sun, W., Roland, K. L. & Curtiss III, R. (2011). Comparison of a regulated delayed antigen synthesis system with *in vivo*-inducible promoters for antigen delivery by live attenuated *Salmonella* vaccines. *Infection and Immunity*, Vol. 79, No. 2, pp. 937-949, ISSN 1098-5522

Waterman, S. R. & Holden, D. W. (2003). Functions and effectors of the *Salmonella* pathogenicity island 2 type III secretion system. *Cellular Microbiology*, Vol. 5, No. 8, pp. 501-511, ISSN 1462-5814

Weber, B., Hasic, M., Chen, C., Wai, S. N. & Milton, D. L. (2009). Type VI secretion modulates quorum sensing and stress response in *Vibrio anguillarum*. *Environmental Microbiology*, Vol. 11, No. 12, pp. 3018-3028, ISSN 1462-2920

Weening, E. H., Barker, J. D., Laarakker, M. C., Humphries, A. D., Tsolis, R. M. & Bäumler, A. J. (2005). The *Salmonella enterica* serotype Typhimurium *lpf, bcf, stb, stc, std,* and *sth* fimbrial operons are required for intestinal persistence in mice. *Infection and Immunity*, Vol. 73, No. 6, pp. 3358-3366, ISSN 0019-9567

White, A. P., Gibson, D. L., Kim, W., Kay, W. W. & Surette, M. G. (2006). Thin aggregative fimbriae and cellulose enhance long-term survival and persistence of *Salmonella*. *Journal of Bacteriology*, Vol. 188, No. 9, pp. 3219-3227, ISSN 0021-9193

Wilson, R. P., Winter, S. E., Spees, A. M. & other authors. (2011). The Vi capsular polysaccharide prevents complement receptor 3-mediated clearance of *Salmonella enterica* serotype Typhi. *Infection and Immunity*, Vol. 79, No. 2, pp. 830-837, ISSN 1098-5522

Winter, S. E., Raffatellu, M., Wilson, R. P., Rüssmann, H. & Bäumler, A. J. (2008). The *Salmonella enterica* serotype Typhi regulator TviA reduces interleukin-8 production in intestinal epithelial cells by repressing flagellin secretion. *Cellular Microbiology*, Vol. 10, No. 1, pp. 247-261, ISSN 1462-5822

World Health Organization, Department of Vaccines and Biologicals. (May 2003). *Background document: the diagnosis, prevention and treatment of typhoid fever.* Geneva, Switzerland, pp. 1-38. Available from www.who.int/entity/vaccine_research/documents/en/typhoid_diagnosis.pdf.

Yu, X. J., Liu, M. & Holden, D. W. (2004). SsaM and SpiC interact and regulate secretion of *Salmonella* pathogenicity island 2 type III secretion system effectors and translocators. *Molecular Microbiology*, Vol. 54, No. 3, pp. 604-619, ISSN 0950-382X

Zavialov, A. V., Kersley, J., Korpela, T., Zav'yalov, V. P., MacIntyre, S. & Knight, S. D. (2002). Donor strand complementation mechanism in the biogenesis of non-pilus systems. *Molecular Microbiology*, Vol. 45, No. 4, pp. 983-995, ISSN 0950-382X

Zenk, S. F., Jantsch, J. & Hensel, M. (2009). Role of *Salmonella enterica* lipopolysaccharide in activation of dendritic cell functions and bacterial containment. *Journal of Immunology*, Vol. 183, No. 4, pp. 2697-2707, ISSN 1550-6606

Zhang, X. L., Tsui, I. S., Yip, C. M., Fung, A. W., Wong, D. K., Dai, X., Yang, Y., Hackett, J. & Morris, C. (2000). *Salmonella enterica* serovar Typhi uses type IVB pili to enter human intestinal epithelial cells. *Infection and Immunity*, Vol. 68, No. 6, pp. 3067-3073, ISSN 0019-9567

Zhao, L., Ezak, T., Li, Z. Y., Kawamura, Y., Hirose, K. & Watanabe, H. (2001). Vi-Suppressed wild strain *Salmonella typhi* cultured in high osmolarity is hyperinvasive toward epithelial cells and destructive of Peyer's patches. *Microbiology and Immunology*, Vol. 45, No. 2, pp. 149-158, ISSN 0385-5600

Zhao, Y., Jansen, R., Gaastra, W., Arkesteijn, G., van der Zeijst, B. A. & van Putten, J. P. (2002). Identification of genes affecting *Salmonella enterica* serovar Enteritidis infection of chicken macrophages. *Infection and Immunity*, Vol. 70, No. 9, pp. 5319-5321, ISSN 0019-9567

Zogaj, X., Nimtz, M., Rohde, M., Bokranz, W. & Römling, U. (2001). The multicellular morphotypes of *Salmonella typhimurium* and *Escherichia coli* produce cellulose as the second component of the extracellular matrix. *Molecular Microbiology*, Vol. 39, No. 6, pp. 1452-1463, ISSN 0950-382X

7

Assembly and Activation of the MotA/B Proton Channel Complex of the Proton-Driven Flagellar Motor of *Salmonella enterica*

Yusuke V. Morimoto and Tohru Minamino
Osaka University
Japan

1. Introduction

Salmonella enterica can swim by rotating multiple flagella, which arise randomly over the cell surface (Fig. 1A). The flagellum consists of at least three parts: the basal body, the hook, and the filament. The basal body is embedded within the cell membranes and acts as a bi-directional rotary motor powered by an electrochemical potential gradient of protons across the cytoplasmic membrane (Fig. 1B, C). The hook and filament extend outwards in the cell exterior. The filament acts as a helical propeller. The hook exists between the basal body and filament and functions as a universal joint to smoothly transmit torque produced by the motor to the filament. When the motors rotate in counterclockwise direction, the cells can swim smoothly. By quick reversal rotation of the motor to clockwise direction, the cells tumble and change their swimming direction to move toward more favorable environments (Fig. 2) (Berg, 2003; Blair, 2003; Minamino et al., 2008).

In *Salmonella enterica*, five flagellar proteins, MotA, MotB, FliG, FliM and FliN, are responsible for torque generation. Two integral membrane proteins, MotA and MotB, form the stator complex of the motor, which functions as a proton-conducting channel to convert an inwardly directed flux of protons through the proton channel into the mechanical work required for motor rotation. Two highly conserved residues, Pro-173 in MotA and Asp-33 in MotB, are involved in the energy coupling mechanism. FliG, FliM and FliN form the C-ring on the cytoplasmic face of the MS ring and are responsible not only for torque generation but also for switching the direction of motor rotation (Berg, 2003; Blair, 2003; Minamino et al., 2008). Torque is generated by sequential electrostatic interactions between MotA and FliG. A high-resolution observation of flagellar motor rotation has revealed a fine stepping motion with 26 steps per revolution, which corresponds to that of FliG subunits in the C-ring (Sowa et al., 2005; Nakamura et al., 2010). The proton conductivity of the MotA/B proton channel complex is proposed to be suppressed by a plug segment of MotB when the MotA/B complex is not assembled around the basal body (Hosking et al., 2006; Morimoto et al., 2010a). Although a low-resolution structure of the entire basal body containing the stator complexes has been visualized in situ by electron cryotomography (Fig. 3A) (Murphy et al., 2006; Liu et al., 2009; Kudryashev et al., 2010), the stators are missing in highly purified flagellar basal bodies presumably due to highly dynamic interactions between the stator and its binding partners in the basal body (Fig. 3B) (Thomas et al., 2001, 2006). In this chapter, we describe our current understanding of how the stator complex is installed into the motor, how its proton conductivity is activated, and how the proton flow through the proton channel is coupled with torque generation.

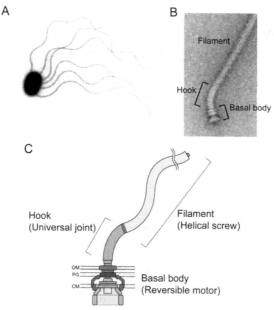

Fig. 1. Bacterial flagella
(A) Electron micrograph of *Salmonella enterica*. Six flagella arise randomly over the cell surface. (B) Electron micrograph of the purified flagellum. (C) Schematic diagrams of the flagellum. The flagellum consists of at least three parts: the basal body, which acts as a reversible rotary motor; the hook, which functions as a universal joint; and the filament, which acts as a helical screw. OM, outer membrane; PG, peptidoglycan layer; CM, cytoplasmic membrane.

Typical swimming pattern of *Salmonella* cells. The cells can swim smoothly when the motors rotate in counterclockwise (CCW) direction, resulting in the formation of a flagellar bundle that produces the thrust. By quick reversal rotation of the motor to clockwise (CW) direction, the bundle is disrupted so that bacteria can tumble and change their swimming direction.

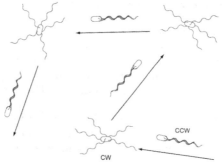

Fig. 2. Bacterial behavior

Assembly and Activation of the MotA/B Proton Channel Complex of the Proton-Driven Flagellar
Motor of Salmonella enterica

149

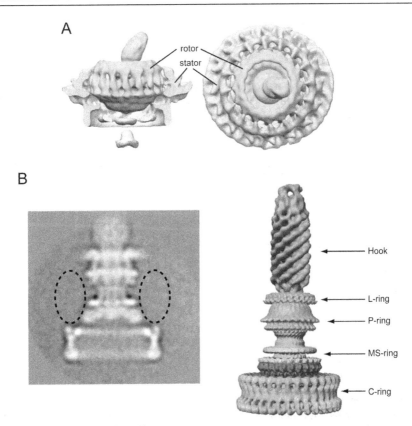

Fig. 3. Structure of bacterial flagellar motor

(A) The entire structure of the *Borrelia burgdorferi* flagellar motor visualized by electron
cryotomography. The segmented structure was color coded as follows: rotor, green; C-ring
and stator, blue; protein export apparatus, gray (Liu et al., 2009). (B) Three-dimensional (3D)
structure of the flagellar basal body isolated from *Salmonella enterica* (Thomas et al., 2006). Left
panel: electron cryomicroscopic image of frozen-hydrated hook-basal body in the side view.
Right panel: 3D density map of the basal body reconstructed from frozen-hydrated particle
images. Dashed circles indicate putative positions of the stator complexes in the motor.

2. MotA/B stator complex acts as a transmembrane proton channel

MotA and MotB are cytoplasmic membrane proteins and form a complex consisting of four
copies of MotA and two copies of MotB (Kojima and Blair, 2004). It has been estimated that
there are at least 11 copies of the MotA/B complex around the flagellar basal body (Reid et al.,
2006). Since MotB has a potential peptidoglycan-binding motif in its C-terminal periplasmic
domain (De Mot and Vanderleyden, 1994), the MotA/B complex is postulated to be anchored
to the peptidoglycan (PG) layer to act as the stator of the motor. The MotA/B complex forms a
transmembrane proton channel that couples an inward-directed proton translocation to torque
generation (Blair and Berg, 1990). However, the molecular mechanism of energy coupling
between proton influx and flagellar motor rotation remains unknown.

The torque-speed relationship of the flagellar motor has been well characterized (Fig. 4)(Chen and Berg, 2000a, b). As the load is decreased, torque decreases gradually up to a certain speed called the knee point and then falls rapidly to zero. The rotation rate of the proton-driven flagellar motor under a given load is proportional to a proton motive force across the cytoplasmic membrane over the entire range of observation (Gabel and Berg, 2003). The plateau torque at high load is dependent on the number of stators while the motor speed near zero load is independent of the stator number (Fig. 4A) (Ryu et al., 2000; Yuan and Berg, 2008). In the high-torque, low-speed regime, both temperature and solvent-isotope effects are small, while those effects are large in the low-torque, high-speed regime. This suggests that the steep decline of torque at high speed is due to a limit in the rate of proton transfer (Fig. 4B) (Chen and Berg, 2000a, b).

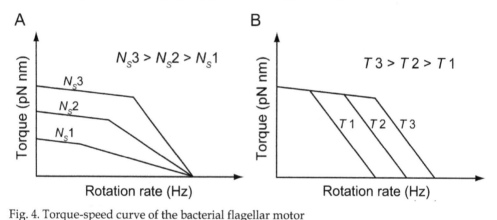

Fig. 4. Torque-speed curve of the bacterial flagellar motor
(A) Effect of the number of functional stators in a motor on the torque-speed relationship. The torque at high load is dependent on the number of stators. The motor speed near zero load is independent of the functional stator number. (Ryu et al., 2000) (B) Effect of temperature on the torque-speed curve. In the low-speed regime, torque is insensitive to changes in temperature. In the high-speed regime, motor speed decreases markedly when temperature is decreased (Chen et al., 2000a).

2.1 Arrangement of transmembrane segments of the MotA/B complex

MotA consists of four transmembrane spans (TMs), two short periplasmic loops and two extensive cytoplasmic regions. MotB consists of an N-terminal cytoplasmic region, one TM and a large periplasmic region containing a putative peptidoglycan-binding motif (Fig. 5A). The four MotA subunits in the stator complex are positioned with their TM3 and TM4 segments adjacent to the dimer of MotB-TM located at the center and their TM1 and TM2 segments in outer positions (Fig. 5B) (Braun et al., 2004). Two conserved charged residues, Arg-90 and Glu-98, in the cytoplasmic loop between TM2 and TM3 are involved in electrostatic interactions with charged residues in the C-terminal domain of FliG to produce torque (Zhou and Blair, 1997; Zhou et al., 1998a). A conserved proline residue of MotA, Pro-173 and Pro-222, are thought to be involved in conformational changes of the stator complex that couple proton influx with torque generation (Braun et al., 1999; Nakamura et al., 2009b). The absolutely conserved and functionally critical aspartic acid residue, Asp-33 of MotB is

located near the cytoplasmic end of its TM and is postulated to be a proton-binding site
(Sharp et al., 1995; Togashi et al., 1997; Zhou et al., 1998b; Che et al., 2008; Morimoto et al.,
2010a). MotB exists as a dimer in the stator complex and these aspartic acid residues are
positioned on the surface of the MotB-TM dimer facing MotA-TMs, suggesting that the
stator complex is likely to have two proton-conducting pathways (Fig. 5B) (Braun and Blair,
2001). Both protonation and deprotonation of this aspartic acid residue cause
conformational changes of the cytoplasmic loop of MotA, which may drive flagellar motor
rotation (Kojima and Blair, 2001; Che et al., 2008).

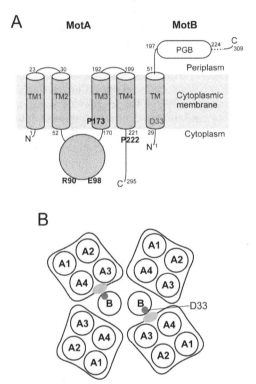

Fig. 5. Topology of MotA and MotB
(A) Cartoon representing the domain organization of MotA and MotB. *Salmonella* MotA and
MotB consist of 295 and 309 amino acid residues, respectively. MotA has four transmembrane
segments (TM1-TM4), two short periplasmic loops and two large cytoplasmic regions. MotB
has an N-terminal cytoplasmic region, one transmembrane segment and a large periplasmic
domain containing a putative peptidoglycan-binding motif (PGB) (Braun et al., 1999, 2004;
Kojima et al., 2004). Arg-90 and Glu-98 of MotA are required for electrostatic interactions with
FliG. Pro-173 and Pro-222 of MotA are proposed to be responsible for conformational changes
of the stator complex that couple proton influx with torque generation. Asp-33 in MotB is
postulated to play an important role in proton translocation. (B) Arrangement of
transmembrane segments of the MotA/B complex, which consists of 4 copies of MotA and 2
copies of MotB. The view is from the periplasmic side of the membrane. The complex has two
proton conducting pathway shown by orange ellipsoids. (Braun et al., 2004).

2.2 Asp-33 of MotB and Pro-173 of MotA are responsible for energy coupling between proton flow and motor rotation

Asp-33 of MotB is a highly conserved and protonatable residue, and is believed to be involved in the proton-relay mechanism (Sharp et al., 1995; Togashi et al., 1997; Zhou et al., 1998b). An increase in the intracellular proton concentration suppresses or even abolishes the flagellar motor function. This suggests that a decrease in the intracellular pH presumably interferes with proton dissociation from Asp-33 to the cytoplasm, thereby slowing the torque generation reaction cycle of the motor (Minamino et al., 2003; Nakamura et al., 2009a). Therefore, it seems likely that both protonation and deprotonation of this Asp play key roles in the torque generation cycle. A D33E mutation in MotB causes ca. 40% reduction in stall torque and a sharp decline in the torque-speed curve with an apparent maximal rotation rate of ca. 20 Hz. This suggests that the D33E mutation not only reduces the proton conductivity significantly but also interferes with an actual torque generation step by the stator-rotor interactions considerably (Che et al., 2008). The stall torque is recovered nearly to the wild-type levels by the suppressor mutations in the transmembrane helices TM2 and TM3 of MotA and in TM and the periplasmic domain of MotB. In contrast, high-speed rotation under low load is still significantly limited even in the presence of these suppressor mutations. These suggest that the second-site mutations recover energy coupling of an inward-directed proton translocation with torque generation but not the maximum proton conductivity (Che et al., 2008).

Pro-173 is highly conserved among MotA orthologs. Since point mutations at Pro-173 affect not only motor function but also proton flow through the MotA/B complex, this prolyl residue seems to play a critical role in the proton-relay mechanism (Braun et al., 2004). *E. coli* *motA*(P173C)/*motB*(D32C) double mutant strain gives a high yield of disulfide-linked MotA/B heterodimers upon oxidation with iodine. This suggests that Pro-173 of MotA is positioned in the proton channel near Asp-32 of MotB, which corresponds to Asp-33 in *Salmonella* MotB (Braun et al., 2004). The *Salmonella* *motA*(P173A) mutant produces stall torque at wild-type levels (Nakamura et al., 2009b). However, the P173A mutation causes a sharp decline in the torque-speed curve with a maximum rotation rate of ca. 25 Hz (Nakamura et al., 2009b). These suggest that the P173A mutation reduces the maximum proton conductivity of the MotA/B complex but not the efficiency of energy coupling between proton translocation and torque generation. Since Pro-173 of MotA is likely to be very close to Asp-33 of MotB in the MotA/B complex structure (Braun et al., 2004), and because Pro is known to induce a kink in α-helix, its replacement with Ala would relax the conformational strain of the MotA/B complex, resulting in a change in protein dynamics to slow down the rates of conformational changes that switch the exposure of Asp-33 to the outside or the inside of the cell. Therefore, Pro-173 is proposed to play an important role in facilitating the resetting of the position of Asp-33 relative to the proton pathway in the conformational change between the two distinct states to facilitate proton translocation when the motor spins at high speed (Nakamura et al., 2009b).

2.3 Role of conserved charged residues in MotA in stator assembly around the motor

Over-expression of MotA inhibits motility of wild-type cells. Neither cell growth, flagellar formation, nor proton motive force across the cytoplasmic membrane is affected by overproduced MotA. However, stall torque is significantly reduced. Since the stall torque is

Assembly and Activation of the MotA/B Proton Channel Complex of the Proton-Driven Flagellar
Motor of Salmonella enterica

153

dependent on the number of functional stators in the motor (Fig. 4), this suggests that MotA occupies the stator-binding sites of the motor and reduces the number of functional stators (Morimoto et al., 2010b). In agreement with this, fluorescent spots of MotA-mCherry have been observed at the same position as those of GFP-FliG even in the absence of MotB (Fig. 6) (Morimoto et al., 2010b). Thus, MotA alone can be installed into a motor.

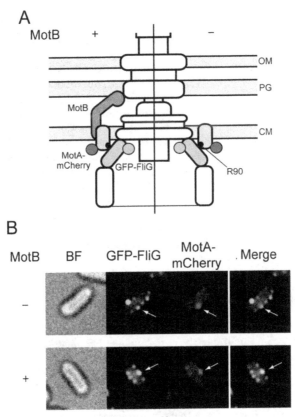

Fig. 6. Co-localization of GFP-FliG and MotA-mCherry
(A) Cartoon representing the basal body and positions of GFP-FliG and MotA-mCherry in the presence (+) or absence (-) of MotB. (B) Bright-field (BF) and fluorescence images of GFP-FliG and MotA-mCherry observed by fluorescence microscopy. The fluorescence images of GFP-FliG (green) and MotA-mCherry (red) are merged in the right panel. (Morimoto et al., 2010b).

Two charged residues of MotA, Arg-90 and Glu-98, are involved in electrostatic interactions with charged residues of FliG for torque generation (Zhou & Blair, 1997; Zhou et al., 1998a). The *motA*(R90E) and *motA*(E98K) alleles are recessive and hence do not exert a negative dominance when their expression levels were the same as those of wild-type cells (Zhou & Blair, 1997; Morimoto et al., 2010b). Interestingly, however, an increase in the expression level of the MotA(R90E)/B complex by more than 10-fold allows 70% of the cells to swim in liquid media (Morimoto et al., 2010b). In agreement with this, the R90E mutation markedly decreases the number and intensity of fluorescent spots of GFP-MotB. Since the loss-of-function phenotype of the *motA*(R90E) and *motA*(E98K) alleles are considerably suppressed by the *fliG*(D289K) and *fliG*(R281V) mutations, respectively (Zhou et al., 1998a), this suggests that the interactions between MotA Arg-90 and FliG Asp-289 and between MotA Glu-98 and FliG Arg-281 are critical not only for torque generation but also for the assembly of the stators into the motor (Morimoto et al., 2010b).

Polar localization of the PomA/B complex of *Vibrio alginolyticus*, which are homologs of the MotA/B complex and acts as the stator of the motor fueled by the sodium motive force across the cytoplasmic membrane, is greatly affected by changes in the external concentration of sodium ions (Fukuoka et al., 2009). This suggests that sodium ions are required not only for torque generation but also for the efficient assembly process of PomA/B. In contrast, neither the D33N nor D33A mutation in *Salmonella* MotB, which abolishes the proton flow through a proton channel, affects stator assembly around the motor, indicating that stator assembly in the proton-driven motor is not dependent on the proton conductivity of the MotA/B complex. In addition, depletion of the proton motive force across the cytoplasmic membrane by a protonophore, carbonyl cyanide *m*-chlorophenylhydrazone does not abolish the subcellular localization of the stator labeled with GFP, indicating that the stators remain to exist around the rotor even in the absence of the proton motive force. This is supported by the finding that the stator actually switches its functional state between the active and inactive ones without detaching from the rotor completely when PMF is largely reduced (Nakamura et al., 2010). Therefore, it is much likely that the assembly of the *Salmonella* MotA/B complex into the motor is not obligatorily linked to the process of the proton translocation through the proton channel of the MotA/B complex.

2.4 MotB$_C$ controls the proton channel activity during stator assembly

The periplasmic domain of MotB (MotB$_C$) consisting of residues 78 through 309 Kojima et al., 2008 contains a putative peptidoglycan-binding (PGB) motif, which shows sequence similarity to other Omp-like proteins (Fig. 7A) (De Mot and Vanderleyden, 1994). Interestingly, the crystal structure of the core domain of *Salmonella* MotB$_C$, which consists of residues 99 through 276 has a typical OmpA-like structure, and shows considerable structural similarities to the C-terminal regions of PAL and RmpM (Fig. 7A) (Kojima et al., 2009). Mutations in the PGB motif of MotB significantly impair motility, indicating that anchoring MotA/B complex to the PG layer is critical for the motor function. Stator resurrection experiments have shown that abrupt drops in the rotation rate occur frequently even in steadily rotating motors (Block and Berg, 1984; Blair and Berg, 1988; Sowa et al., 2005). Consistently, fluorescent photo-bleaching studies of GFP-fused MotB have shown a rapid exchange and turnover of the stator complexes between the membrane pool and the basal body (Leake et al., 2006). These results suggest that the association between MotA/B and its target site on the basal body are highly dynamic.

MotB$_C$ forms homo-dimer in solution (Fig. 7B) (Kojima et al., 2008, 2009). Over-expression of MotB$_C$ inhibits motility of wild-type cells when expressed in the periplasm. An in-frame deletion of resiudes 197-210 not only inhibits dimerization of MotB$_C$ but also significantly reduces its inhibitory effect on wild-type motility. These results suggest that dimerization of MotB$_C$ is responsible for the proper targeting and stable anchoring of the MotA/B complex to the putative stator binding sites of the basal body (Kojima et al., 2008). Recently, site-directed disulfide crosslinking experiments have shown an interaction between MotB$_C$ and the P ring of the basal body (Hizukuri et al., 2010).

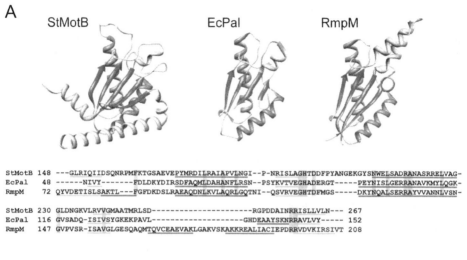

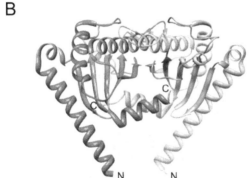

```
StMotB 148 ---GLRIQIIDSQNRPMFKTGSAEVEPYMRDILRAIAPVLNGI--P-NRISLAGHTDDFPYANGEKGYSNWELSADRANASRRELVAG-
EcPal   48 ------NIVY-------FDLDKYDIRSDFAQMLDAHANFLRSN--PSYKVTVEGHADERGT-----PEYNISLGERRANAVKMYLQGK-
RmpM    72 QYVDETISLSAKTL----FGFDKDSLRAEAQDNLKVLAQRLGQTNI--QSVRVEGHTDFMGS-----DKYNQALSERRAYVVANNLVSN-

StMotB 230 GLDNGKVLRVVGMAATMRLSD--------------------RGPDDAINRRISLLVLN--- 267
EcPal  116 GVSADQ-ISIVSYGKEKPAVL--------------------GHDEAAYSKNRRAVLVY----- 152
RmpM   147 GVPVSR-ISAVGLGESQAQMTQVCEAEVAKLGAKVSKAKKREALIACIEPDRRVDVKIRSIVT 208
```

Fig. 7. Crystal structure of the C-terminal domain of MotB
(A) Comparison of various OmpA-like domains. Cα ribbon diagrams of three OmpA-like
domains and their structure-based sequence alignment. StMotB, *Salmonella typhimurium*
MotB$_{C2}$ (2zvy); EcPal, *Escherichia coli* Pal (1oap); RmpM, the C-terminal domain of RmpM
from *Neisseria meningitidis* (1r1m). Conserved secondary structural elements are colored:
orange, α-helix; purple, β-strand. Residues highlighted in light blue in the aligned sequences
are conserved in these three proteins. Regions of secondary structures are indicated below
the corresponding sequences: red line, α-helix; green line, β-strand. (B) Cα ribbon
representations of the MotB$_{C2}$ dimer. The two subunits are shown in blue and orange.
(Kojima et al., 2009)

Cell growth is not impaired significantly by over-expression of the *E. coli* MotA/B complex
(Stolz and Berg, 1994). In contrast, the cell growth is severely impaired by co-expression of
MotA with a MotB-TetA chimera protein, in which the first 60 residues of MotB are fused to
a 50 residue sequence encoded by an open reading frame in the *tetA* gene (Blair and Berg,
1990). Since the replacement of Asp-32 to other residues in the MotB-TetA chimera protein
has been shown to suppress proton leakage caused by co-overproduction of MotA with the

MotB-TetA chimera protein. Therefore, it has been suggested that the proton conductivity of the MotA/B complex is suppressed prior to stator assembly around a motor. An in-frame deletion of residues 51-70 or 52-71 within the periplasmic domain of *E. coli* MotB or *Salmonella* MotB, respectively, a region, which is highly conserved among the MotB orthologs, causes considerable proton leakage, thereby arresting cell growth (Hosking et al., 2006; Morimoto et al., 2010a). This suggests that the deleted region of MotB acts as a plug segment that inserts into a proton channel within the cytoplasmic membrane to prevent from premature proton translocation through the channel before the association with a motor and that upon stator assembly in the motor, the plug leaves the channel, allowing the stator to conduct protons (Fig. 8). Interestingly, however, cell growth is not significantly impaired by in-frame deletion of residues 51-100 in *Salmonella* MotB, which contains the plug segment (Muramoto and Macnab, 1998; Kojima et al., 2009). This indicates that a proton channel of the MotA/B(Δ51-100) complex is not formed prior to its assembly around the rotor although the plug is missing. The introduction of the L119P or L119E substitution into MotB(Δ51-100) causes growth impairment due to significant proton leakage (Kojima et al., 2009; Morimoto et al., 2010a). Therefore, it is likely that some other region within the periplasmic domain of MotB also regulates proper formation of the proton channel during stator assembly.

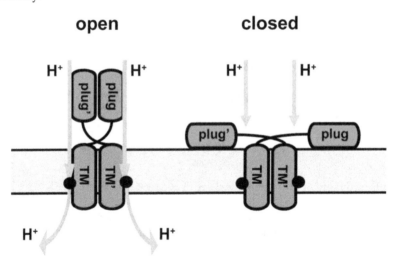

Fig. 8. Schematic diagram of the plug region of MotB

Cartoon representing the position of the plug in the open and closed states of the MotA/B proton channel. Only the MotB TMs and plugs are shown, in a view parallel with the plane of the cytoplasmic membrane. Asp-33 in MotB is shown as a black dot. Blue arrows indicate the proton flow through proton channels. In the open state, the plugs leave the membrane and associate with each other via their hydrophobic faces to hold the proton channel open. In the closed state, the plugs insert into the cell membrane parallel with its periplasmic face and interfere with channel formation (Hosking et al., 2006).

MotB(Δ51-100) is still functional, indicating that MotB(Δ51-100) can form a functional stator complex along with MotA (Muramoto and Macnab, 1998). Since the distance between the

Assembly and Activation of the MotA/B Proton Channel Complex of the Proton-Driven Flagellar
Motor of Salmonella enterica

157

surface of the hydrophobic core layer of the cytoplasmic membrane and that of the PG layer is about 100 Å, and the crystal structure of the core domain of MotB$_C$ is only about 50 Å tall, the crystal structure of the core domain of *Salmonella* MotB$_C$ is so small that MotB(Δ51-100) cannot reach the PG layer if connected directly to the transmembrane helix by the deletion of residues 51-100. Therefore, a large conformational change would be required for the PGB sites on the top surface of MotB$_C$ to reach the PG layer. Since a specific interaction between a cytoplasmic loop between TM-2 and TM-3 in MotA and FliG is required for stator assembly around a rotor of the motor, it is possible that the interaction between MotA and FliG may trigger conformational changes in MotB$_C$ that open the proton channel and allow the stator to be anchored to the PG layer (Fig. 9).

MotA and MotB are colored blue and orange, respectively. Arg-90 of MotA, which is required for stator assembly into the motor, is shown as a black dot. An interaction between MotA and FliG may trigger conformational changes in MotB$_C$ that open the proton channel and allow the stator to be anchored to the PG layer (Kojima et al., 2009; Morimoto et al., 2010b).

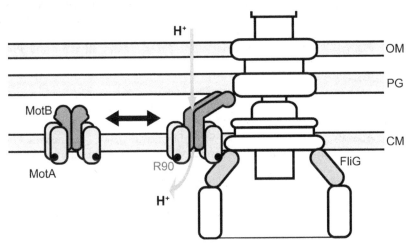

Fig. 9. Model for stator assembly into the motor

3. Conclusion

The MotA/B complex, which is composed of four copies of MotA and two copies of MotB, acts as a proton channel to couple proton flow to torque generation. The crystal structure of the C-terminal periplasmic domain of MotB (MotB$_C$) shows significant structure similarities to other PGB domains such as the C-terminal regions of PAL and RmpM, suggesting that an association of MotB$_C$ with the PG layer anchors the MotA/B complex to be the stator around the rotor. Interestingly, the MotA/B complex does not always associate with a motor during flagellar motor rotation, suggesting that the association of the MotA/B complex to its target sites on the basal body is highly dynamic. A plug segment in the periplasmic domain of MotB prevents the proton channel from leaking protons into the cytoplasm when the MotA/B complex is not assembled into the motor. An electrostatic interaction between MotA and FliG is required not only for the efficient assembly of the MotA/B complexes around the rotor but also for the proton channel formation to conducts protons coupled with torque generation.

4. Acknowledgment

We acknowledge K. Namba for continuous support and encouragement and N. Kami-ike and Y-S. Che and S. Nakamura for helpful discussions. Y.V.M. is a research fellow of the Japan Society for the Promotion of Science. This work has been supported in part by Grants-in-Aid for Scientific Research from the Ministry of Education, Culture, Sports, Science and Technology of Japan to Y.V.M. and T.M.

5. References

Berg, H.C. (2003) The rotary motor of bacterial flagella. *Annu Rev Biochem* 72: 19–54.

Blair, D.F., & Berg, H.C. (1988) Restoration of torque in defective flagellar motors. *Science* 242: 1678–1681.

Blair, D.F., & Berg, H.C. (1990) The MotA protein of *E. coli* is a proton-conducting component of the flagellar motor. *Cell* 60: 439–449.

Blair, D. F. (2003) Flagellar movement driven by proton translocation. *FEBS Lett* 545: 86–95.

Block, S.M., & Berg, H.C. (1984) Successive incorporation of force-generating units in the bacterial rotary motor. *Nature* 309: 470–472.

Braun, T. & Blair, D.F. (2001) Targeted Disulfide Cross-Linking of the MotB Protein of *Escherichia coli*: Evidence for Two H+ Channels in the Stator Complex. *Biochemistry* 40: 13051–13059.

Braun, T., Poulson, S., Gully, J.B., Empey, J.C., Van Way, S., Putnam, A., & Blair, D.F. (1999) Function of proline residues of MotA in torque generation by the flagellar motor of Escherichia coli. *J Bacteriol* 181: 3542–3551.

Braun, T.F., Al-Mawasawi, L.Q., Kojima, S. & Blair, D.F. (2004) Arrangement of core membrane segments in the MotA/MotB protein-channel complex of *Escherichia coli*. *Biochemistry* 43: 35–45.

Che, Y-S., Nakamura, S., Kojima S., Kami-ike, N., Namba, K. & Minamino, T. (2008) Suppressor analysis of the MotB(D33E) mutation to probe the bacterial flagellar motor dynamics coupled with proton translocation. *J Bacteriol* 190: 6660–6667.

Chen, X. & Berg, H. C. (2000a) Torque-speed relationship of the flagellar motor of *Escherichia coli*. *Biophys J* 78: 1036–1041.

Chen, X. & Berg, H. C. (2000b) Solvent-isotope and pH effects on flagellar rotation in *Escherichia coli*. *Biophys J* 78: 2280–2284.

De Mot, R. & Vanderleyden, J. (1994) The C-terminal sequence conservation between OmpA-related outer membrane proteins and MotB suggests a common function in both gram-positive and gram-negative bacteria, possibly in the interaction of these domains with peptidoglycan. *Mol Microbiol* 12: 333–334.

DeRosier, D. (2006) Bacterial Flagellum: Visualizing the Complete Machine In Situ. *Current Biology* 16: 928–930.

Fukuoka, H., Wada, T., Kojima, S., Ishijima, A. & Homma M. (2009) Sodium-dependent dynamic assembly of membrane complexes in sodium-driven flagellar motors. *Mol Microbiol* 71: 825–835.

Gabel, C.V. & Berg, H.C. (2003) The speed of the flagellar rotary motor of *Escherichia coli* varies linearly with protonmotive force. *Proc Natl Acad Sci USA* 100: 8748–8751.

Hizukuri, Y., Kojima, S. & Homma, M. (2010) Disulfide cross-linking between the stator and the bearing components in the bacterial flagellar motor. *J Biochem* 148: 309–318.

Assembly and Activation of the MotA/B Proton Channel Complex of the Proton-Driven Flagellar
Motor of Salmonella enterica

159

Hosking, E.R., Vogt, C., Bakker, E.P. & Manson, M.D. (2006) The *Escherichia coli* MotAB proton channel unplugged. *J Mol Biol* 364: 921–937.

Kihara, M. & Macnab, R.M. (1981) Cytoplasmic pH mediates pH taxis and weak-acid repellent taxis of bacteria. *J Bacteriol* 145: 1209–1221.

Kojima, S. & Blair, D.F. (2001) Conformational change in the stator of the bacterial flagellar motor. *Biochemistry* 40: 13041–13050.

Kojima, S. & Blair, D.F. (2004) Solubilization and purification of the MotA/MotB complex of *Escherichia coli. Biochemistry* 43: 26–34.

Kojima, S., Furukawa, Y., Matsunami, H., Minamino, T. & Namba, K. (2008) Characterization of the periplasmic domain of MotB and implications for its role in the stator assembly of the bacterial flagellar motor. *J Bacteriol* 190: 2259–2266.

Kojima, S., Imada, K., Sakuma, M., Sudo, Y., Kojima, C., Minamino, T., Homma, M. & Namba, K. (2009) Stator assembly and activation mechanism of the flagellar motor by the periplasmic region of MotB. *Mol Microbiol* 73: 710–718.

Kudryashev, M., Cyrklaff, M., Wallich, R., Baumeister, W. & Frischknecht, F. (2010) Distinct in situ structures of the Borrelia flagellar motor. *J Struct Biol* 169:54–61.

Leake, M.C., Chandler, J.H., Wadhams, G.H., Bai, F., Berry, R.M. & Armitage, J.P. (2006) Stoichiometry and turnover in single, functioning membrane protein complexes. *Nature* 443: 355–358.

Liu, J., Lin, T., Botkin, D.J., McCrum, E., Winkler, H. & Norris, S.J. (2009) Intact Flagellar Motor of Borrelia burgdorferi Revealed by Cryo-Electron Tomography: Evidence for Stator Ring Curvature and Rotor/C-Ring Assembly Flexion. *J Bacteriol* 191: 5026–5036.

Minamino, T., Imae, Y., Oosawa, F., Kobayashi, Y. & Oosawa, K. (2003) Effect of intracellular pH on the rotational speed of bacterial flagellar motors. *J Bacteriol* 185: 1190–1194.

Minamino, T., Imada, K. & Namba, K. (2008) Molecular motors of the bacterial flagella. *Curr Opin Struct Biol* 18:693–701.

Morimoto, Y.V., Che, Y-.S., Minamino, T. & Namba, K. (2010a) Proton-conductivity assay of plugged and unplugged MotA/B proton channel by cytoplasmic pHluorin expressed in *Salmonella. FEBS lett* 584: 1268–1272.

Morimoto, Y.V., Nakamura, S., Kami-ike, N., Namba, K. & Minamino, T. (2010b) Charged residues in the cytoplasmic loop of MotA are required for stator assembly into the bacterial flagellar motor. *Mol Microbiol* 78: 1117–1129

Muramoto, K. & Macnab, R.M. (1998) Deletion analysis of MotA and MotB, components of the force-generating unit in the flagellar motor of *Salmonella. Mol Microbiol* 29: 1191–202.

Murphy, G.E., Leadbetter, J.R. & Jensen, G.J. (2006) In situ structure of the complete Treponema primitia flagellar motor. *Nature* 442: 1062–1064.

Nakamura, S., Kami-ike, N., Yokota, P.J., Kudo, S., Minamino, T. & Namba, K. (2009a) Effect of intracellular pH on the torque-speed relationship of bacterial proton-driven flagellar motor. *J Mol Biol* 386: 332–338.

Nakamura, S., Morimoto, Y.V., Kami-ike, N., Minamino, T. & Namba, K. (2009b) Role of a conserved prolyl residue (Pro-173) of MotA in the mechanochemical reaction cycle of the proton-driven flagellar motor of Salmonella. *J Mol Biol* 393: 300–307.

Nakamura, S., Kami-ike, N., Yokota, J.P., Minamino, T. & Namba, K. (2010) Evidence for symmetry in the elementary process of bidirectional torque generation by the bacterial flagellar motor. *Proc Natl Acad Sci USA* 107: 17616–17620.

Reid, S.W., Leake, M.C., Chandler, J.H., Lo, C.J., Armitage, J.P. & Berry, R.M. (2006) The maximum number of torque-generating units in the flagellar motor of *Escherichia coli* is at least 11. *Proc Natl Acad Sci USA* 103: 8066–8071.

Ryu, W.S., Berry, R.M. & Berg, H.C. (2000) Torque-generating units of the flagellar motor of *Escherichia coli* have a high duty ratio. *Nature* 403: 444–447.

Sharp, L.L. Zhou, J. & Blair, D.F. (1995) Tryptophan-scanning mutagenesis of MotB, an integral membrane protein essential for flagellar rotation in *Escherichia coli*, *Biochemistry* 34, 9166–9171.

Sowa, Y., Rowe, A.D., Leake, M.C., Yakushi, T., Homma, M., Ishijima, A. & Berry, R.M. (2005) Direct observation of steps in rotation of the bacterial flagellar motor. *Nature* 437: 916–919.

Stolz, B. & Berg, H.C. (1991) Evidence for interactions between MotA and MotB, torque-generating elements of the flagellar motor of *Escherichia coli*. *J Bacteriol* 173: 7033–7037.

Thomas, D., Morgan, D.G. & DeRosier, D.J. (2001) Structures of bacterial flagellar motors from two FliF-FliG gene fusion mutants. *J Bacteriol* 183: 6404–6412.

Thomas, D.R., Francis, N.R., Chen, X. & DeRosier, D.J. (2006) The three-dimensional structure of the flagellar rotor from a clockwise-locked mutant of *Salmonella enterica* serovar Typhimurium. *J Bacteriol* 188: 7039–7048.

Togashi, F., Yamaguchi, S., Kihara M., Aizawa, S.-I. & Macnab, R.M. (1997) An extreme clockwise switch bias mutation in *fliG* of *Salmonella typhimurium* and its suppression by slow-motile mutations in *motA* and *motB*. *J Bacteriol* 179: 2994–3003.

Yuan, J., & Berg, H.C. (2008) Resurrection of the flagellar rotary motor near zero load. *Proc Natl Acad Sci USA* 105: 1182–1185.

Zhou, J. & Blair, D.F. (1997) Residues of the cytoplasmic domain of MotA essential for torque generation in the bacterial flagellar motor. *J Mol Biol* 273: 428–439.

Zhou, J., Lloyd, S. A. & Blair, D.F. (1998a) Electrostatic interactions between rotor and stator in the bacterial flagellar motor. *Proc Natl Acad Sci USA* 95: 6436–6441.

Zhou, J., Sharp, L.L., Tang, H.L., Lloyd, S.A. & Blair, D.F. (1998b) Function of protonatable residues in the flagellar motor of *Escherichia coli*: a critical role for Asp32 of MotB. *J Bacteriol* 180: 2729–2735.

8

Comprehending a Molecular Conundrum: Functional Studies of Ribosomal Protein Mutants from *Salmonella enterica* Serovar Typhimurium

Christina Tobin Kåhrström[1], Dan I. Andersson[1] and Suparna Sanyal[2,*]
[1]Department of Medical Biochemistry and Microbiology
[2]Department of Cell and Molecular Biology
Uppsala University
Sweden

1. Introduction

Of all the molecules present in the living organism, proteins are the most abundant and are often described as the most fundamental. This is a reasonable assertion, considering the enormous diversity of crucial cellular functions that are performed by proteins, which range from structural support and signal transduction to enzyme catalysis and immune defence. Given the vital nature of their function, efficient and orderly production of proteins is essential for cell fitness and survival.

In every living cell, protein synthesis or translation as it is otherwise known, is performed by complex, dynamic ribonucleoprotein particles known as ribosomes. Conversion of the genetic code into biologically active proteins is an elaborate process that relies on the availability of mRNA template, as well as amino acid substrates that are delivered to the ribosome attached to tRNA molecules. In addition, the ribosome is assisted by a number of protein factors, known as "translation factors", which serve to streamline and accelerate various sub-steps of the process. Having said that, however, all phases of the translation cycle can be performed in the absence of such factors, albeit slow, so the ribosome can be regarded as the core of translation. It manages to manufacture proteins at an incredible speed (up to 20 amino acids per second) and with very few errors (one in every 1,000 – 10,000 codons deciphered) (Kramer & Farabaugh, 2007). In *Escherichia coli (E. coli)*, ribosomes account for up to 50% of the dry mass of the cell and considering their central role in gene expression, it is not surprising that the cell devotes up to 40% of its total energy to ribosome production during rapid growth (Nierhaus, 1991; Nierhaus, 2006). The immensity of the task of the ribosome is adequately reflected in the sophistication of its structure, where the modern ribosome represents one of the largest and most complex cellular machines. Despite substantial progress in terms of genetic, biochemical and structural techniques, designed to delineate the mechanism of action of the ribosome, many aspects of protein synthesis and ribosome function remain to be elucidated.

* Corresponding Author

1.1 Bacterial ribosomes – Structure and function

Ribosomes are universally conserved and are composed of ribosomal RNA (rRNA) and ribosomal proteins (r-proteins) in all organisms. Ribosomal RNA is the major structural component of all cytoplasmic ribosomes, where it accounts for approximately two thirds of the total mass. The remaining third consists of a large number of mostly small, basic ribosomal proteins (r-proteins) that are scattered over the surface of the structure. The simplest ribosome, that of the bacterial cell, totals approximately 2.5 MDa and is over 200 Å in width (Williamson, 2009), establishing it as the largest enzymatic particle in the cell. The number of ribosomes in the cell can reach up to tens of thousands in order to cope with the increased demand for protein synthesis during periods of rapid cell division (Bremer & Dennis, 1996).

All ribosomes consist of two subunits of unequal size. In prokaryotes the smaller of the two subunits is known as the 30S and the larger subunit is known as the 50S, which are named based on their coefficient of sedimentation. The subunits exist as free, separated molecules in the cell cytoplasm when inactive but associate to form a translation-competent 70S ribosome upon the initiation of protein synthesis. The 30S is composed of one strand of 16S rRNA (1,542 nucleotides) and 21 r-proteins. As expected the 50S contains more rRNA (23S and 5S containing 2,904 and 120 nucleotides, respectively) and more protein (33 r-proteins). For many years, the detailed structure of the ribosome was elusive due to limitations of crystallographic techniques, which was compounded by the tremendous size of the particle. However, at the turn of the century, crystallographic images of ribosomal subunits at the atomic level appeared and marked a gigantic leap forward in our understanding of the structural mysteries (Ban et al., 2000; Wimberly et al., 2000). Since then detailed images of

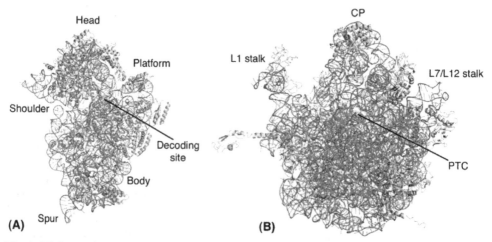

Fig. 1. **High-resolution crystal structures of the ribosomal subunits.** Tertiary structures of the bacterial 30S (A) and 50S (B) viewed from the inter-subunit face. Ribosomal RNA depicted in blue and r-proteins as green ribbons. Abbreviations are as follows: CP, central protuberance; PTC, peptidyl-transferase centre. The image was generated in PyMol (www.pymol.org) from the coordinates of the crystal structure of the *T. thermophilus* ribosome (PDB ID 2J00 & 2J01)(Selmer et al., 2006).

Comprehending a Molecular Conundrum: Functional Studies of Ribosomal Protein Mutants from
Salmonella enterica Serovar Typhimurium

163

the full ribosome with translation ligands bound have been generated and to date the highest resolution crystal structure available is that of the *Thermus thermophilus (T. thermophilus)* 70S ribosome (Figure 1) (Selmer et al., 2006).

As clearly illustrated by the crystal structure, the two subunits are markedly different in their overall shape. The morphology of each subunit is an inherent feature of the folds adopted by each rRNA species and gives rise to a rigid large subunit and a more flexible small subunit. The 50S is a compact, compressed particle consisting of a hemispherical base with three conspicuous protuberances designated the L1 stalk, the central protuberance (CP) and the L7/L12 stalk (Figure 1B). The two rRNA strands (23S and 5S rRNAs) are highly interlaced to form a dense rRNA mesh. In contrast, the 30S is not only smaller in size, but is also less compact. The four secondary structure domains of the 16S rRNA correlate well to distinct regions of the three dimensional structure such as the head, body and platform (Figure 1A). The r-proteins are spatially dispersed at the periphery and back of the subunits, remote from the main functional sites such as the decoding centre, peptidyl-transferase centre (PTC) and the inter-subunit face. With this realisation, the controversy surrounding the basis of ribosomal enzymatic activity was finally laid to rest since the structures established that the PTC was devoid of protein and the ribosome was quickly recognised as a ribozyme (Nissen et al., 2000). In addition, high-resolution structures of the 30S revealed that decoding was primarily mediated by dynamic rearrangements of the 16S rRNA upon cognate tRNA binding at the ribosomal A-site (Ogle et al., 2003). Together, these studies propelled rRNA into the spotlight, securing its superior position as the major mediator of ribosomal function.

1.2 The ribosome at work

A detailed description of bacterial translation is beyond the scope of this text, however a brief overview is presented to facilitate a clear understanding of the experiments that will be described later. The process can be conceptually divided into four major phases as described below.

1.2.1 Initiation

Initiation begins on the 30S subunit and requires the binding of an mRNA template that carries the blueprint for the protein product. Next, selective and rapid binding of the unique, initiator tRNA (fMet-tRNA^{fMet}) to the 30S subunit takes place. This event is coordinated by the action of three initiation factors (IF1, IF2 and IF3), the principal one being the G-protein factor IF2, forming the 30S preinitiation complex. The 50S subunit is recruited and the initiation factors are released to form what is known as a 70S initiation complex (IC) that is now ready to proceed to the next phase.

1.2.2 Elongation

During elongation, the polypeptide grows as a result of the processive addition of amino acids that are delivered to the 70S ribosome as aminoacyl-tRNAs in a ternary complex with EF-Tu and GTP. Upon acceptance of the cognate substrate, rapid peptidyl transfer occurs. The tRNA-mRNA complex is re-positioned by a distance of exactly one codon to facilitate further addition of aminoacyl-tRNAs and this translocation step is stimulated by GTP hydrolysis on EF-G. Amino-acyl tRNA selection, peptidyl transfer and translocation are repeated until a stop codon appears at the A-site.

1.2.3 Termination

In eubacteria, two release factor proteins (RF1 and RF2) recognize stop codons in a semi-specific manner and activate release of the nascent protein from the P-site tRNA to terminate translation. A third release factor RF-3, removes and recycles RF1 and RF2.

1.2.4 Ribosome recycling

During the final phase, ribosome recycling factor (RRF), in combination with EF-G are required to split the 70S into its subunits. The deacylated tRNA attached to the P-site of the 30S is ejected upon IF3 binding, the 30S can slide along the mRNA to re-initiate translation or the mRNA is also released. This event leaves the ribosome and its factors available to participate in further rounds of translation.

1.3 Ribosomal proteins

Despite the predominance of rRNA, the bacterial ribosome does contain an impressive array of 54 individual r-proteins. In *E. coli* and *Salmonella enterica* serovar Typhimurium, 21 proteins are associated with the small subunit (S proteins) and 33 with the large subunit (L proteins) (Wilson & Nierhaus, 2005). Among eukaryotes, the number of r-proteins increases by approximately 20-30 (Doudna & Rath, 2002). In bacteria, all r-proteins, with the exception of L7/L12, are present as a single copy per ribosome.

The overall abundance of rRNA and its direct involvement in the major tasks of the ribosome (decoding and peptidyl-transfer) have allowed the r-proteins to evade direct scrutiny and their contribution to translation is largely ill-defined. Furthermore, many of these proteins tend to be highly conserved, which forcefully calls our attention to questions regarding their functional roles (Lecompte et al., 2002). In general, they are considered important for the maintenance and stability of the overall structure of the ribosome (Cech, 2000). This assumption is largely due to their extensive contact with rRNA where many of the proteins contain extended projections that insert into the busy network of rRNA helices, prompting the idea that they probably promote or may even facilitate correct folding of the rRNA into its active conformation. However, few studies have directly addressed this issue and we are in the dark regarding the functional roles of most of the proteins. With this in mind, our research has focused on ascertaining the requirement and involvement of particular r-proteins for optimal cell growth and ribosome activity. We have posed the following questions:

1. If rRNA is responsible for the basic function of the ribosome then what, if any, are the selective advantages offered by the r-proteins? In other words, are they all actually necessary for ribosome function?
2. Is it possible to create mutants that completely lack a r-protein and mutants that carry a non-native version of the protein from another species?
3. What are the physiological consequences of such mutations with respect to cell fitness and ribosome function?

Comprehending a Molecular Conundrum: Functional Studies of Ribosomal Protein Mutants from
Salmonella enterica Serovar Typhimurium

165

4. If such mutations confer fitness costs (i.e. slower growth), is it possible to compensate for such costs?
5. What are the mechanisms of fitness compensation that evolve in response to removal and replacement of r-proteins?

By addressing such questions we are ultimately interested in probing the complexity and robustness of the translation apparatus. To approach some useful answers to these questions, we have used *Salmonella enterica* serovar Typhimurium LT2 (hereafter referred to as *S. typhimurium*) as a model organism and introduced deletions of the genes encoding the r-proteins S20 and L1 and replaced S20, L1 and L17 with orthologous counterparts from close and phylogenetically distant relatives, including the archaeon *Sulfolobus acidocaldarius* and the eukaryote *Saccharomyces cerevisiae*.

2. Using *S. typhimurium* as a model in the ribosome field

E. coli has long served as a model organism for genetic and biochemical studies of bacterial ribosome function. However, due to the co-linearity of their genomes, *S. typhimurium* represents a convenient alternative for such studies and is preferable in our case due to a more favourable and better characterized genetic background for the purposes of genetic suppression studies (McClelland et al., 2001) (see below). Although highly similar in terms of ribosomal gene content, *S. typhimurium* does possess some unique features that are described below.

2.1 Primary sequences of rRNA and r-proteins

In terms of the basic building blocks of the ribosome, rRNA and r-proteins, the genes encoding these components are generally highly conserved between *E. coli* and *S. typhimurium*, a reflection of their close phylogenetic relationship. In both organisms, there are seven independent rRNA operons (*rrnA, rrnB, rrnC, rrnD, rrnE, rrnG* and *rrnH*), each encoding a single copy of the 16S, 23S and 5S rRNAs. The genes encoding the 16S and 5S rRNAs show strong similarities; however, more variation is detected between the genes encoding the 23S rRNA species (*rrl* genes). In particular, the primary sequences of *rrl* genes in *S. typhimurium* are between 90 and 110 nucleotides longer due to the presence of so-called intervening sequences (IVSs) that are removed during maturation (Burgin et al., 1990). Despite this discrepancy in the length of their primary 23S rRNA sequences, the mature, core sequence in *S. typhimurium* is nearly identical (97%) to that of *E. coli* (Burgin et al., 1990).

In terms of the r-proteins, a high degree of conservation is also observed upon alignment of the amino acid sequences of *S. typhimurium* LT2 proteins and *E. coli* K-12 MG1655 proteins. For the 21 r-proteins associated with the small subunit, a median similarity of 99.2% is detected with an average identity of 98.9% and a standard deviation of 0.9%. Similarly, for the 33 r-proteins associated with the large subunit, the median amino acid identity is 98.9% with an average similarity of 97.7% and a standard deviation of 3.6%. From these calculations, we can conclude that there is slightly more variation in the amino acid sequences of large subunit proteins when comparing the primary sequences of *S. typhimurium* LT2 and *E. coli* K-12 MG1655. However, overall, a high degree of conservation exists, which is not surprising given the close phylogenetic relationship.

2.2 rRNA processing during ribosome biogenesis

During transcription of the rRNA operons, the primary transcripts are processed to mature products by means of RNase cleavage and chemical modification (Kaczanowska & Ryden-Aulin, 2007). The vast majority of studies that have examined these processing events have looked at the *E. coli* system only, however it is expected that *rrn* processing in close relatives such as *S. typhimurium* should be highly similar.

Primary *rrn* transcripts of *E. coli* are initially substrates for the endoribonuclease RNase III that catalyzes separation of the rRNAs into pre-16S (17S), pre-23S and pre-5S (9S) species (Nierhaus, 2006). Final trimming of the 5' and 3' ends of each transcript involves a number of exonucleases and occurs in the context of pre-ribosomal particles (Deutscher, 2009). In some genera of bacteria, including *S. typhimurium*, mature 23S rRNA is not fully intact but is fragmented during processing. Experiments have revealed that the fragments arise due to RNase III-mediated removal of intervening sequences (Burgin et al., 1990). Although the resulting fragments of 23S rRNA are not re-ligated they are fully functional. Some heterogeneity exists between the seven copies of 23S rRNA genes in *S. typhimurium* LT2 with regard to the presence of these additional sequences. In the case of *rrlG* and *rrlH*, an insertion of approximately 110 nucleotides is observed in helix-25 (located at nucleotide position 550, approximately) and *rrlA, rrlB, rrlC, rrlD, rrlE* and *rrlH* contain an insertion of approximately 90 nucleotides in helix-45 (located at nucleotide position 1170, approximately). Interestingly, ribosomes of a *S. typhimurium* mutant lacking RNase III remain functional, implying that fragmentation is not required for the activity of *S. typhimurium* ribosomes (Mattatall & Sanderson, 1998). This would also suggest that it is unlikely that these extra sequences evolved to support proper functioning of *S. typhimurium* ribosomes. Their intermittent distribution among diverse genera of bacteria would suggest that they may have been acquired by horizontal gene transfer (Mattatall & Sanderson, 1996).

2.3 Unusual sucrose gradient profile of *Salmonella* ribosomes

Analysis of ribosome function *in vitro* requires the extraction of large quantities of intact ribosomes from the cell using a standard sucrose density ultracentrifugation protocol (Johansson et al., 2008). Typically, this involves preparing a pure cell lysate that is passed through two sucrose cushions to remove cell debris. Ribosomal particles (70S, 50S and 30S) are subsequently separated on a final sucrose gradient (10 – 50%) based on differences in their densities. The standard protocols available are optimised for *E. coli* and pure fractions of each ribosomal particle are easily obtainable. The high degree of similarity between *S. typhimurium* and *E. coli* might, *a priori*, lead one to expect a similar profile when the protocol is applied to *S. typhimurium*. However, in our hands, we have consistently observed a striking difference.

Looking at a typical, large-scale 70S purification profile from *E. coli* K12 strain MRE600 (Figure 2A), clean, distinct peaks are observed for each ribosomal particle, as indicated. In contrast, the 70S peak from wild-type *S. typhimurium* LT2 contains a distinct shoulder (Figure 2B). Upon examination of the protein and rRNA content of the second, smaller peak, we observed that this contains 30S particles only, indicating that the 70S and 50S peaks coincide when ribosomes are prepared from *S. typhimurium* on a large scale. This is a consistent feature of the wild-type strain, as well as all r-protein mutant strains that we have

Comprehending a Molecular Conundrum: Functional Studies of Ribosomal Protein Mutants from
Salmonella enterica Serovar Typhimurium

167

worked with. At the present time, there is no definitive explanation for these observations but the unusual co-migration of 70S and 50S species may be related to 23S rRNA processing differences between the species, or, 70S ribosomes of *S. typhimurium* may be composed of loosely-coupled subunits that require a higher magnesium concentration for stabilization.

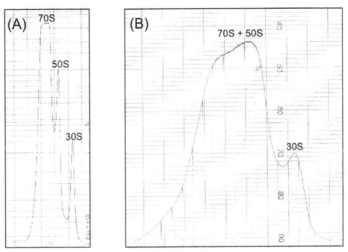

Fig. 2. **Sucrose gradient ultracentrifugation profiles obtained during 70S ribosome purification.** (A) Profile obtained from *E. coli* MRE600 showing three separate peaks for 70S, 50S and 30S particles. (B) Profile obtained from wild-type *S. typhimurium* showing a distinctive shouldered peak containing a mixture of 70S and 50S particles followed by a 30S peak.

2.4 Advantages of using *S. typhimurium* as a model

E. coli is traditionally regarded as the first choice model organism for studies of ribosome function. However, it has recently become clear that the most frequently used K-12 strain has mutations in components that function in translation. Mutations in the genes *prmB* (encoding class-I release factor RF2) and *rpsG* (encoding ribosomal protein S7) represent two well-characterized examples. O' Connor et al. have reported that in all K-12 strains of *E. coli*, RF2 carries a threonine residue at position 246 as opposed to alanine, which is typically found at this position in the majority of bacterial RF2 proteins (O'Connor & Gregory, 2011). In the case of *rpsG*, a point mutation alters its canonical UGA stop codon to a leucine codon (Schaub & Hayes, 2011). Thus, the S7 protein of all K-12 strains carries an extension of 23 amino acids at its C-terminus. This protein is essential and particularly important during 30S assembly. Both of these mutations were identified as being partially responsible for the fitness cost and ribosomal perturbations associated with an *rluD* null mutation.

These studies serve to highlight the importance of using a model organism with a clean genetic background. The presence of underlying, undetected mutations complicates the matter of defining the mechanism and cause of observed phenotypes associated with a mutation of interest due to the unavoidable likelihood of synergistic and epistatic interactions. Our research has concentrated on the effects of r-protein removal and replacement in terms of bacterial fitness and ribosome function with a strong focus on suppression genetics. So, although *E. coli* K-12 is undoubtedly the most well-studied model organism, *S. typhimurium* represents a more suitable and reliable model for the purposes of our studies.

3. Isolation and characterisation of ribosomal protein mutants

The scarcity of studies designed to examine the distinct functional roles of r-proteins, despite their high level of conservation in all three phylogenetic kingdoms, certainly warrants further experimentation aimed at understanding their contributions to cell fitness and ribosome activity. To address this issue, genetically defined *S. typhimurium* mutants were constructed in which the target proteins were either completely removed (S20 and L1) from the chromosome or replaced (S20, L1 and L17) on the chromosome with orthologous counterparts from other species.

3.1 Previous studies of r-protein mutants

The majority of the early research on mutants lacking r-proteins was initiated by the work of Dabbs and colleagues. Beginning in the mid-70s, various antibiotic selection systems were developed for the isolation of *E. coli* mutants with alterations in r-proteins (Dabbs & Wittman, 1976; Dabbs, 1978). Most of the mutants that lacked a r-protein were isolated by chemically mutagenizing antibiotic sensitive cells, selecting for antibiotic dependence followed by selection for reversion of the antibiotic dependent phenotype (Dabbs, 1979). Using such methods, *E. coli* strains were isolated lacking one or more of 16 individual r-proteins (S1, S6, S9, S13, S17, S20, L1, L11, L15, L19, L24, L27, L28, L29, L30 and L33) (Dabbs, 1991). In all of these studies, absence of each protein was only verified on the protein level by means of two-dimensional gel electrophoresis and immunological methods using antibodies specific for each missing protein. The mutants exhibited a conditional lethal phenotype and were varied in terms of growth properties. These studies were certainly pioneering in terms of showing the non-essential nature of certain r-proteins but follow-up studies designed to determine the nature of the fitness costs and ribosomal defects were hampered by the chemical mutagenesis and selection methods used for the isolation of mutants. These strains were subject to mutagenesis on a genome-wide scale and were therefore not fully defined genetically, which creates difficulty in specifically linking the observed phenotypes to absence of the protein. In the past few years, much progress has been made in the techniques used for the manipulation of bacterial genetics (Datsenko & Wanner, 2000; Sharan et al., 2009). These methods have made it possible to introduce precise deletions of targeted genes and a number of research groups have been successful in obtaining r-protein null mutants (Wower et al., 1998; Cukras & Green, 2005; Bubunenko et al., 2007; Korepanov et al., 2007). In a recent work the number of the L7/L12 proteins have been varied by deletion of the binding site of one L7/L12 dimer on the L10 protein (Mandava et al., 2011). The characterisation of such mutants has advanced our current knowledge of the functions of some of the r-proteins; unfortunately, however, these mutants are all limited to the species *E. coli*.

Attempts to produce functionally active hybrid ribosomes harboring non-native r-proteins have previously been undertaken (Liu et al., 1989; Giese & Subramanian, 1991). The majority of these studies only examined functional activity in terms of polysome formation. However, highly divergent proteins such as S18 from rye chloroplast, that shares only 35% amino acid identity with *E. coli* S18, was incorporated into 30-40% of *E. coli* monosomes and polysomes upon expression from a plasmid (Weglohner et al., 1997). The ability of these hybrid ribosomes to form polysomes indicates that they were capable of forming functional complexes, although the extent of their functionality during specific steps of translation was not examined. In the same study, L12 from *Arabidopsis* chloroplasts was also incorporated into ribosomes but these hybrid ribosomes failed to form functional polysomes *in vivo* and were inactive in poly(U)-dependent poly(Phe) synthesis *in vitro*. In another study, the *E. coli* proteins of the GTPase-associated centre (L10, L11 and L7/L12) were replaced with their rat counterparts *in vitro* (Uchiumi et al., 2002). Although incorporated into the *E. coli* ribosome, eukaryotic elongation factors were required to support protein synthesis *in vitro*. So although divergent r-proteins can be assembled into the bacterial ribosome, proteins with defined functions, such as interactions with bacterial specific translation factors, lack function and do not preserve protein synthesis.

3.2 Construction of r-protein mutants

In our studies, mutants of *S. typhimurium* that completely lacked an r-protein, and mutants in which the native r-protein gene was replaced with closely related as well as phylogenetically distant orthologues were created. Due to its simplicity and precision, lambda red recombineering was used for the construction of all mutants (Datsenko & Wanner, 2000). For removal of S20 and L1, a kanamycin resistance cassette was PCR amplified using primers with 50 nucleotide 5' tails homologous to the flanking sequence of the target gene (*rplA* and *rpsT*). Electroporation of the linear DNA product into the cell resulted in substitution of the r-protein gene with the selectable kanamycin resistance cassette to create mutants lacking S20 (Tobin et al., 2010) and L1 (Figure 3A). A similar procedure was used for the generation of S20, L1 and L17 replacement mutants, however in this case the PCR fragment consisted of the open reading frame (ORF) of the orthologous replacement gene and an adjacent kanamycin resistance cassette (Lind et al., 2010). Homologous recombination resulted in replacement of the target r-protein ORF (*rpsT, rplA* and *rplQ*) with its orthologous ORF and the kanamycin resistance cassette (Figure 3B). Congenic wild-type control strains were also constructed and carried the kanamycin cassette adjacent to the native r-protein ORFs. In the case of both types of constructions, the native promoter and terminator sequences were preserved. Using this technique, individual mutants lacking the r-proteins S20 (ΔS20) and L1 (ΔL1) as well as replacement mutants carrying heterologous versions of S20, L1 and L17 were created.

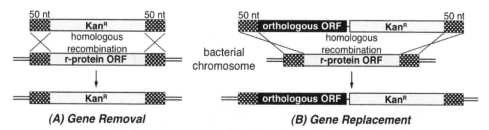

(A) Gene Removal   *(B) Gene Replacement*

Fig. 3. **Schematic diagram representing the technique employed to generate r-protein mutants.** The recombining fragments in (A) and (B) were designed with 50 nucleotides of homology to the flanking sequences of the r-protein target genes to facilitate homologous recombination and replacement on the chromosome.

3.3 Quantification of fitness costs

In a laboratory setting, determining exponential growth rate is the most straightforward method for assaying fitness. To quantify the physiological effect of r-protein removal and replacement, we measured the doubling time of the wild-type and ribosomal mutant strains using a Bioscreen C analyzer. Relative fitness values were calculated as the growth rate of each mutant normalized to that of the wild-type reference strain used in each experiment (set to 1.0). Although relatively easy to execute, this measurement fails to evaluate fitness over the entire growth cycle and is unable to detect very small differences in fitness (sensitivity of ~ 3%). A more labour intensive method that overcomes such limitations is pair-wise competition experiments where the test strain and wild-type are genetically tagged and compete over many generations (sensitivity of ~ 0.3%). This method was used as a more sensitive fitness parameter in some of our experiments.

3.4 Compensatory evolution

Removal and replacement of highly conserved r-proteins, that function within a central component of the cell, constitute deleterious mutations and confer fitness costs, i.e. slower growth. Because such mutants grow more slowly and therefore produce fewer progeny, they will eventually be out-grown by competitors of higher fitness and eliminated from the population. However, if the mutant can persist in the population, it has the potential to develop further mutations that suppress or compensate the fitness cost of the initial deleterious mutation. Such compensatory mutations (denoted CM) can restore fitness back to the wild-type level and effectively neutralize the deleterious effect of the initial mutation (deletion or replacement of a r-protein). Alternatively, the fitness cost may be only partially relieved by the CM, which increases fitness but not to the extent of wild-type fitness (Figure 4A). Here, we are interested in knowing if and by what mechanisms the fitness costs of L1 removal and S20, L1 and L17 replacement can be compensated. This is not only helpful in the sense that it may provide clues to the functional roles of the proteins, but it also provides a unique opportunity to reveal novel co-operative mechanisms in ribosome function.

The experiment is designed to optimize purifying selection, which reduces genetic diversity so that the population stabilizes on the most favourable (fittest) genotype. During the

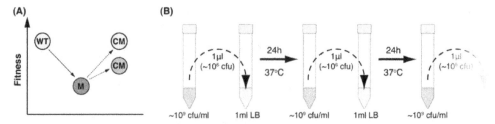

Fig. 4. **Compensatory evolution by serial passage.** (A) Schematic illustration of the principle of compensatory evolution. A deleterious mutation (M) is introduced into a wild-type population (WT) and reduces fitness. Compensatory mutations (CM) that reduce the fitness cost of the initial deleterious mutation can restore fitness completely (yellow) or partially restore fitness (green). (B) Schematic illustration of the serial passage experiment used for the isolation of compensated mutants.

evolutionary process, CMs that reduce the fitness cost conferred by loss or replacement of the r-protein become more common in successive generations. Independent lineages of the mutants are serially passaged in liquid growth medium (LB) for a number of generations. To maintain continuous population growth and to ensure that all genetic variants are represented during the entire evolutionary process, large bottlenecks of $\sim 10^6$ cells are used (Figure 4B). This experimental design maximises the potential to fix the compensated mutant of highest fitness. When mutants of higher fitness constitute the majority of the population, the experiment is terminated and a single colony is examined for the presence of candidate compensatory mutations that increase fitness. To verify that the identified mutations are responsible for improving fitness, genetic reconstructions are performed and fitness is quantified.

3.5 *In vivo* and *in vitro* analysis of ribosomal function

For a full appreciation of the functional roles of deleted r-proteins, both *in vivo* and *in vitro* techniques were employed to compare the activity of mutant ribosomes with their wild-type counterparts.

3.5.1 *In vivo* system

Synthesis of an easily detectable reporter protein was used here to measure the rate of polypeptide elongation and ribosomal misreading of stop codons and frameshifting. Polypeptide elongation rates of exponentially growing cultures were measured by determining the exact time taken to produce the first detectable β-galactosidase molecule following IPTG induction (Miller, 1992). As *S. typhimurium* does not harbor the *lac* operon, the normal inducible *lac* operon was supplied on an F' factor and was transferred to strains by means of conjugation. This method was used to determine the rate of polypeptide elongation in the strains lacking r-proteins S20 and L1.

The synthesis of β-galactosidase was also used to determine the frequency of stop codon read-through and frameshifting of an L1-depleted ribosome. In this instance, strains carried an F' factor in which the *lacI* and *lacZ* genes are fused for constitutive expression of β-galactosidase. However, the ribosome must read-through a premature stop codon (UGA_{189} or UAG_{220}) or a +1 or -1 frameshift in *lacI* to produce an active molecule of the enzyme, the activity of which is detected as described above. A read-through value is calculated based on the amount of enzyme produced by the mutated *lacIZ* construct relative to that produced in a congenic strain harbouring a non-mutated *lacIZ* construct (Bjorkman et al., 1999). A second assay, based on luciferase expression, was also employed to determine if L1 depleted ribosomes were prone to general misreading of sense codons (Kramer & Farabaugh, 2007).

3.5.2 *In vitro* system

The translation cycle is composed of a large number of intermediate sub-steps that proceed extremely fast *in vivo*. Thus, in order to specifically examine defined steps in translation, analysis of ribosome function in a controlled *in vitro* translation system is absolutely necessary. For the characterization of S20- and L1-depleted ribosomes, we have employed a reconstituted cell-free translation system that is optimized to meet *in vivo* expectations in so far as it allows measurement of *in vitro* reactions in the millisecond range (Ehrenberg et al., 1989). Ribosomal particles and all other translation components are prepared to high purity according to standard protocols. Additionally, all experiments are performed in polymix buffer that mimics the pH and complex ionic conditions of the bacterial cell (Pettersson & Kurland, 1980). Using this system, r-protein depleted ribosomes can be accurately compared with their wild-type counterparts in all steps of the translation cycle. Although the system was developed for *E. coli*, it has previously been shown that the system can be applied to ribosomes from *S. typhimurium* (Tubulekas et al., 1991).

4. R-protein deletion mutants

Genetically defined mutants of *S. typhimurium*, lacking individual r-proteins (ΔS20 and ΔL1), represent powerful tools for the analysis of the function of these proteins in the context of translation and can also reveal their potential roles in other cellular processes. Below follows a summary of the fitness costs, compensatory mechanisms and the *in vivo* and *in vitro* analysis of the mutants lacking S20 and L1.

4.1 Removal of S20 and L1 confers substantial fitness costs

Loss of small subunit protein S20 reduced exponential growth rate approximately three-fold relative to the wild-type under standard growth conditions (LB broth at 37°C), corresponding to a relative fitness value of 0.33 (Table 1) (Tobin et al., 2010). Upon deletion of large subunit protein L1, the generation time of the mutant was reduced two-fold compared to wild-type under standard growth conditions (Table 1). This fitness cost was even more pronounced during growth at lower temperatures, indicative of a cold-sensitive phenotype. In the case of both mutants, complete restoration of the wild-type growth rate was observed upon induced expression of the deleted gene from an arabinose-inducible plasmid, verifying that the fitness costs were specifically attributable to loss of each respective protein (Figure 5A and B).

Strain	Growth Temperature	Relative Growth Rate
Wild-Type	n/a	1.0
ΔS20	37°C	0.33
ΔL1	30°C	0.36
ΔL1	37°C	0.52
ΔL1	44°C	0.58

Table 1. **Fitness costs associated with removal of S20 and L1.** The generation time of the wild-type strain is set to 1.0 and is used to normalize the growth rate of each mutant at the temperatures specified.

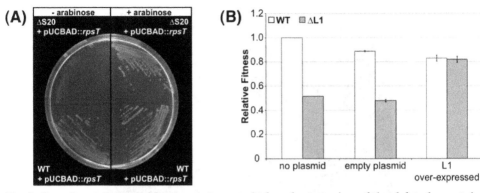

Fig. 5. **Complementation of fitness costs upon induced expression of the deleted r-protein gene from an inducible plasmid.** (A) In the absence of the arabinose inducer, the ΔS20 strain grows very poorly relative to the wild-type (left-hand side of image). In the presence of arabinose, expression of the *rpsT* gene is induced and growth of the mutant matches that of wild-type (right-hand side of image) (Tobin et al., 2010). (B) Growth rates of the ΔL1 mutant relative to the wild-type. In the presence of the arabinose inducer, *rplA* is expressed and fitness of the mutant matches that of wild-type. Error bars represent standard error of the mean.

4.2 Compensatory mutations that mitigate the costs of L1 deletion

Twelve independent lineages of the ΔL1 mutant were evolved for approximately 300 generations to determine if and by what mechanisms compensation could be achieved. Upon isolation of faster-growing mutants, four were chosen at random and subject to whole-genome sequencing to identify candidate compensatory mutations responsible for suppressing the fitness cost associated with loss of L1. A single nucleotide substitution in the *rplB* gene and a small deletion in the *rplS* gene were detected. These mutations resulted in the amino acid substitution E194K in large subunit protein L2 (encoded by *rplB*) and loss of the Leu codon at position 100 of L19 (referred to as *rplS* ΔL100) (Figure 6A). Using a different ΔL1 ancestral strain, the amino acid substitution P102L in the *rplN* gene (encoding large subunit protein L14) was also identified (Figure 6A). Strain reconstructions confirmed that each individual mutation was necessary and sufficient for fitness improvement and all three increased growth rate to a similar extent. In any combination, all three compensatory mutations demonstrated negative epistasis in the ΔL1 background. In the wild-type strain, each individual mutation reduced fitness and this became even more pronounced when the mutations were combined (Figure 6B).

4.3 Phenotypes of previous L1 deletion mutants

Two independent *E. coli* knockout mutants of L1 were previously isolated as spontaneous revertants of chemically mutagenized spectinomycin- and kasugamycin-dependent strains (Dabbs, 1977; Dabbs, 1980). Examination of the activity of ribosomes purified from these strains suggested that absence of L1 reduced the *in vitro* production of polypeptide by 50%, corresponding to an approximate 50% reduction in growth rate (Subramanian & Dabbs, 1980). To probe further into the ribosomal deficiencies, a follow-up study examined the

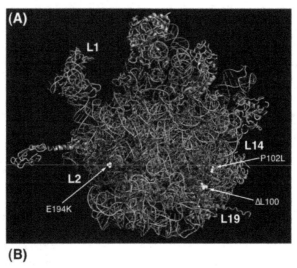

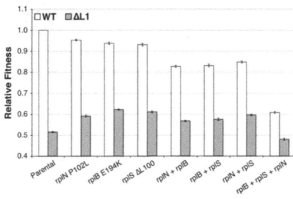

Fig. 6. **Compensation for the fitness costs of L1 deletion.** (A) Crystal structure of the *T. thermophilus* 50S subunit showing the localization of compensatory mutations (yellow spheres) in the r-proteins L2 (red), L14 (purple) and L19 (orange). Position of the L1 protein is shown in green. All other rRNA and protein residues are depicted in grey. The image was created in PyMol (www.pymol.org) using the coordinates of the *T. thermophilus* 50S subunit structure (PDB ID 2J01) (Selmer et al., 2006). (B) Genetic reconstructions and fitness effects of compensatory mutations in the ΔL1 (green bars) and wild-type (white bars) backgrounds. Relative fitness was calculated as the growth rate of each strain as a fraction of the wild-type growth rate. Error bars represent standard error of the mean.

activity of L1-depleted ribosomes using a number of partial reactions. It was found that both the binding of N-acetyl-Phe-tRNA to the P-site and stimulation of EF-G dependent GTP hydrolysis in the presence of tRNA and mRNA were significantly compromised with these ribosomes (Sander, 1983). Another L1 deletion mutant was isolated in an independent study as a spontaneous revertant of a chemically mutagenized erythromycin-dependent strain (Wild, 1988). This strain produced an excess of free ribosomal subunits and fewer 70S

Comprehending a Molecular Conundrum: Functional Studies of Ribosomal Protein Mutants from
Salmonella enterica Serovar Typhimurium

175

ribosomes when the sedimentation profile of ribosomes was examined. In addition, a genetically defined *rplA* deletion strain has been constructed in *E. coli* (Baba et al., 2006), but at the time of writing no details of the phenotype of this mutant are known.

4.3.1 Misreading phenotype

The proposed involvement of L1 in regulating release of the E-site tRNA and the coupling of decoding accuracy at the A-site to occupation of the E-site (Wilson & Nierhaus, 2006) prompted us to investigate the decoding properties of the L1-depleted ribosome. By means of β-galactosidase assays (described in 3.5.1), the level of frameshifting and stop codon readthrough was measured *in vivo*. The only significant difference between the wild-type and mutant was detected with respect to UGA readthrough, which was increased approximately three to five-fold in the case of the ΔL1 mutant, independent of codon context. Misreading of sense codons was also examined although no significant differences were detected, suggesting that L1 plays a minor role, at most, in decoding at the A-site. Furthermore, none of the identified compensatory mutations altered readthrough of UGA in the ΔL1 background.

4.3.2 Polysome analysis

It was previously shown that L1 has RNA chaperone activity (Ameres et al., 2007), so a role for L1 in ribosome biogenesis is possible. To investigate the distribution of different ribosomal particles in the cell, polysome profiles were examined and compared to the profile obtained from wild-type cells (Figure 7). The most striking observation of the mutant

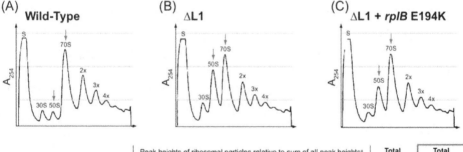

Strain	Peak heights of ribosomal particles relative to sum of all peak heights[a]						Total polysomes[b]	Total 70S particles[c]
	30S	50S	70S	2x	3x	4x		
Wild-Type	0.08	0.07	0.37	0.22	0.15	0.11	0.48	0.85
ΔL1	0.11	0.23	0.29	0.16	0.11	0.09	0.36	0.65
ΔL1+ *rplB* E194K (CM)	0.09	0.19	0.32	0.18	0.12	0.09	0.39	0.71

Fig. 7. **Polysome profiles obtained following immediate translational arrest in log-phase.** Distribution of ribosomal particles in the wild-type (A), ΔL1 mutant (B) and the ΔL1 mutant with the compensatory mutation (CM) in *rplB* (encoding L2). Each peak corresponds to each different particle as labelled and polysomes are labelled as 2x, 3x and 4x. [a] The table shows average peak heights for each particle relative to the the sum of all peak heights and were based on at least three independent profiles. [b] Total polysomes were calculated as the sum of 2x, 3x and 4x peak heights. [c] Total 70S particles were calculated as the sum of 70S, 2x, 3x and 4x peak heights.

profile (Figure 7B) was the large excess of free 50S subunits, which was increased three-fold compared to the wild-type (Figure 7A). Furthermore, this increase in free subunits was biased specifically towards the 50S as the amount of free 30S was only slightly increased. Thus, removal of L1 leads to an imbalance in formation of the ribosomal subunits. This increase in free subunits occurred at the expense of functional 70S complexes, which were reduced from 0.85 in the wild-type to 0.65 in the mutant. However, the ratio of 70S monosomes to polysomes was the same for the mutant and wild-type, indicating that those 70S ribosomes that were formed could translate as efficiently as the wild-type. Upon overexpression of L1, these anomalies in ribosomal particle partitioning were removed, specifically identifying loss of L1 as the cause of the altered profile. The reduction of 70S complexes as monosomes and polysomes is likely to be responsible for the slow growth rate of the ΔL1 mutant as the translational capacity of the strain is likely to be substantially compromised. Interestingly, the E194K amino acid substitution in L2 conferred a small but reproducible reduction in the proportion of free ribosomal subunits and a consequent increase in the fraction of active 70S complexes (Figure 7C).

4.4 L1 plays a major role in ribosome biogenesis

A number of *in vivo* and *in vitro* methods were used in this study to evaluate the role of L1 in ribosome function. Based on the results of these assays, we propose that large subunit protein L1 is crucial for promoting balanced formation of the ribosomal subunits in *S. typhimurium*. This defect is further supported by the observed cold sensitivity of the ΔL1 mutant, a feature that is often associated with aberrant ribosome biogenesis (Dammel & Noller, 1993; Bubunenko et al., 2006). It is not yet clear why the production of free ribosomal subunits is disproportionate, but a few alternative explanations are feasible. L1 has known RNA chaperone activity so the protein may be directly involved in biogenesis of the small subunit; however, no evidence is currently available to support this suggestion. The effect of L1 deletion on 30S production may be more indirect. It is conceivable that in the absence of L1, 50S subunits are more prone to misassembly than the wild-type. This, in turn, may perturb 30S assembly via sequestering of accessory factors required for the biogenesis of both subunits, for example. Alternatively, if the 30S subunits lack properly assembled association partners, final maturation of the 30S may be impaired and/or the 30S subunits may be more susceptible to the action of degradative RNases as previously suggested (Deutscher, 2009). Although only one of the three confirmed compensatory mutations (*rplB* E194K) displayed a detectable improvement in ribosomal particle partitioning (Figure 7C), the remaining two (*rplN* P102L and *rplS* ΔLeu) occur in proteins directly involved in formation of bridging contacts required for 70S complex formation (Selmer et al., 2006). Thus, it would appear as though these two compensatory mutations may alter subunit interactions and possibly increase 70S production. Furthermore, additional compensatory mutations in the genes *rpsM* (encoding S13 that forms bridges B1a and B1b with the 50S subunit) and *engA* (encoding the 50S biogenesis factor Der) provide more evidence that L1 plays an important role in ribosome biogenesis.

Along with helices 76-78 of the 23S rRNA, L1 occupies a highly mobile region of the 50S subunit known as the L1 stalk. Crystal structures of the 50S and 70S revealed that this stalk can adopt both an "open" and "closed" conformation, suggestive of an underlying functional role in the release of deacylated tRNA from the E-site thereby alluding to a role for this region in the translocation step of polypeptide elongation (Harms et al., 2001; Yusupov et al., 2001). These observations sparked a number of structural-based investigations aimed at investigating L1 stalk dynamics which all indicate that this region is involved in modulating decaylated tRNA transit and release during translocation (Valle et al., 2003; Fei et al., 2008; Munro et al., 2010; Trabuco et al., 2010). Thus, to determine the activity of L1 depleted ribosomes in the elongation step of translation, a number of *in vivo* and *in vitro* assays are currently underway.

4.5 Phenotype of S20 deletion mutant

Knockout mutants of S20 have been isolated previously as antibiotic dependent revertants and as a suppressor of nonsense codons (Dabbs, 1978; Dabbs, 1979; Ryden-Aulin et al., 1993). Both types of mutants displayed initiation defects in terms of initiator tRNA binding (Gotz et al., 1990) and subunit association (Gotz et al., 1989; Ryden-Aulin et al., 1993). In another study, an S20 knockout mutant was obtained as a suppressor of erythromycin dependence and produced precursor 30S particles and fewer 70S complexes, as measured by density ultracentrifugation (Wild, 1988). More recently, the recombineering approach has been used to isolate genetically defined, in-frame deletions of *rpsT* in *E. coli*, but data to describe the phenotypic effects of these deletions is currently unavailable (Baba et al., 2006; Bubunenko et al., 2007).

4.5.1 Polypeptide formation *in vivo*

The rate at which the ΔS20 mutant forms polypeptide *in vivo* was compared to the wild-type strain using the β-galactosidase assay. Similar to the results of the ΔL1 mutant, no defect in the rate of polypeptide chain elongation was detected, however a reproducible reduction in the rate of accumulation of synthesized protein was evident (Tobin et al., 2010). Considering that S20 is a designated 30S subunit protein and given the results of previous studies, we hypothesized that the reduced rate of protein accumulation was likely the result of impaired initiation. A reduction in the frequency of mRNA initiation would directly decrease the number of rounds of translation per mRNA and reduce the yield of protein production. Thus, the activity of S20-depleted 30S subunits were compared to those obtained from the wild-type, using a number of *in vitro* methods designed to measure each distinct step of translation initiation.

4.5.2 Reduced mRNA binding *in vitro*

The rate of template binding was measured using a radioactively labeled mRNA with a strong Shine-Dalgarno (SD) sequence. In both the presence and absence of initiator tRNA and the initiation factors, the rate of mRNA binding to ΔS20 30S subunits was reduced approximately 3.5-fold and maximum binding required 40 minutes compared to only 10 minutes for the wild-type. Since similar rates of mRNA binding were observed with and without the initiation ligands, a primary impairment in mRNA binding could be concluded. Reconstitution with purified S20 restored the wild-type mRNA binding phenotype and established that defective mRNA association was a direct consequence of S20 removal (Figure 8A). In addition, titration of mRNA revealed that a substantial fraction (~ 25%) of the S20-depleted 30S subunits were completely inactive in the binding of mRNA (Figure 8B). Since the rate of fMet-tRNA^{fMet} binding emulated that of mRNA and was therefore used to detect the level of mRNA bound to 30S subunits in this assay, the possibility of unstable initiator tRNA binding could not be entirely excluded as the underlying cause of incomplete ΔS20 30S occupancy with mRNA. Thus the stability of initiator tRNA binding was measured by its rate of dissociation from 30S pre-initiation complexes. The results demonstrated that dissociation of fMet-tRNA^{fMet} occurred at the same rate for the mutant and wild-type. Hence, we could conclude that reduced initiator tRNA binding was a secondary, consequential effect of a primary impairment in mRNA binding (Tobin et al., 2010).

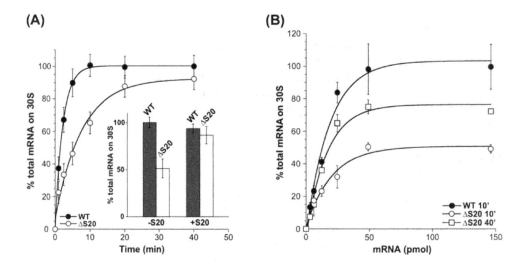

Fig. 8. **Activity of wild-type and S20-depleted 30S subunits in mRNA binding.** (A) The rate of mRNA binding to the mutant 30S subunit is reduced by a factor of approximately 3.5. Upon reconstitution with purified S20 protein, this impairment is removed (insert). (B) Upon titration with mRNA, a considerable fraction of the mutant 30S subunits failed to associate with mRNA. (Tobin et al., 2010)

4.5.3 Defective subunit association

The kinetics of 50S association to wild-type and S20-depleted 30S subunits was also measured in a stopped-flow instrument using light scattering. This assay demonstrated that naked ΔS20 30S subunits were severely impaired in association with wild-type 50S subunits and no improvement was observed when the 50S was added in excess (Figure 9A). However, a pronounced improvement (from ≤ 20% to ~ 40%) in the association capacity of the mutant 30S was observed upon pre-incubation with mRNA, initiator tRNA and the initiation factors. Similar to the results of mRNA binding, subunit association was further improved upon extension of the ΔS20 30S pre-incubation time with initiation components to 40 minutes (Figure 9B) (Tobin et al., 2010).

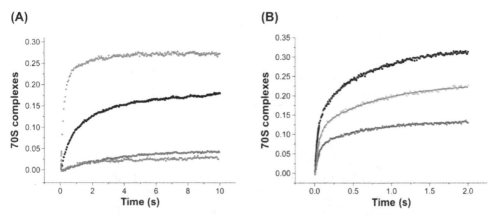

Fig. 9. **Kinetics of subunit association measured by light scattering.** (A) In the absence of
the initiation ligands and when the ratio of 30S to 50S is 1:1, very few mutant 70S complexes
are formed (red circles) compared to wild-type (black squares). The addition of 5x more 50S
to the reaction (wild-type, grey squares; ΔS20, blue circles), did not increase 70S formation in
the case of the mutant. (B) Association kinetics in the presence of mRNA, initiator tRNA, IF1
and IF2 following pre-incubation for 10 minutes (wild-type, black squares; ΔS20, red circles)
and 40 minutes (ΔS20, green circles). (Tobin et al., 2010)

4.6 S20 is required for correct structural positioning of h44

Defects in the rate and extent of mRNA binding, as well as the poor association capacity of
S20-depleted 30S subunits are likely to account for the reduced rate of protein accumulation
observed *in vivo* and the prolonged generation time of the ΔS20 mutant. These impairments
were puzzling as we could not reconcile them with the topographical location of S20 in
mature 30S particles. The protein (Figure 10) is located at the base of the body of the 30S
subunit, distal to the mRNA binding channel and subunit interface. However S20 has been
shown to interact with helix 44 of the 16S rRNA during 30S assembly and in mature 30S
particles (Brodersen et al., 2002; Dutca & Culver, 2008). High-resolution crystal structures
have shown that this helix stretches across the entire length of the 30S body and forms part
of the A- and P- sites where mRNA binding occurs (Schluenzen et al., 2000) and it also forms
many bridging contacts with the 50S subunit (Selmer et al., 2006). Hence, upon removal of
S20, h44 most likely adopts a suboptimal structural position that impairs the binding of
mRNA and inhibits association with the 50S subunit (Tobin et al., 2010). Since prolonged
incubation with the initiation components concealed these defects to some degree, it is likely
that these 30S ligands promote a structural rearrangement of h44 that permits more stable
binding of mRNA and facilitates docking of the 50S subunit.

5. R-protein replacement mutants

Besides the knowledge that can be gained regarding the essentiality and putative roles of r-
proteins by examining genetically defined deletion mutants, r-protein replacement studies
offer a unique opportunity to ask some more general questions. Using lambda red
recombineering, the ORFs of S20, L1 and L17 were replaced with both closely related and

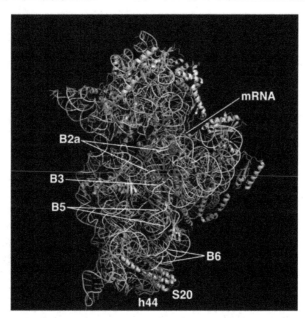

Fig. 10. **Structure of the 30S subunit highlighting the proposed involvement of S20 in mRNA binding and subunit association.** S20 (green ribbon) interacts with and possibly stabilizes h44 (red) in an orientation that is optimized for mRNA (orange; A-site, purple; P-site, cyan) binding and interactions with the 50S subunit (bridges B2a, B3, B5 and B6 depicted in yellow) (Tobin et al., 2010). All other rRNA and protein residues are depicted in grey. The image was created in PyMol (www.pymol.org) from the coordinates of the *T. thermophilus* 30S subunit (PDB ID 2J00) (Selmer et al., 2006).

highly divergent orthologues on the chromosome of *S. typhimurium* (Lind et al., 2010). Such hybrid mutants address fundamental evolutionary questions regarding translational robustness and the plasticity of the *bone fide* ribosomal components. Highly sensitive fitness measurements were performed to directly examine the costs of replacement and to indirectly determine the functional capacity of hybrid ribosomes. In addition, compensatory evolution of replacement mutants with large fitness costs revealed a general mechanism for suppressing loss of functionality when a divergent protein is imported into the ribosome.

5.1 Modest fitness costs upon r-protein replacement

Considering the highly conserved nature of r-proteins and their involvement in one of the key processes of gene expression, their removal was expected to confer substantial reductions in cell fitness. The physiological effects of r-protein replacement however, was more difficult to predict, although it was reasonable to expect relatively weaker fitness costs, even if the non-native protein lacked a high degree of homology to the host protein. The non-native replacement proteins ranged from close relatives, such as those from other proteobacteria *E. coli* and *Klebsiella pneumoniae*, to phylogenetically distant orthologues such as those from the eukaryote *Saccharomyces cerevisiae* and the archaeon *Sulfolobus acidocaldarius* (Lind et al., 2010). Exponential growth rates were measured in four different media where the doubling time of the wild-type strain varied from 20 – 120 minutes and the growth rates of hybrid strains were expressed relative to this (Figure 11A showing L1 replacement mutants only). In the case of the S20 replacement mutants, all strains had higher fitness than the null mutant, irrespective of the growth medium used and even when amino acid identity fell to only 32%. Similarly, the fitness of the majority of L1 replacement mutants was only modestly reduced relative to wild-type, although more substantial costs were evident when the amino acid identity of the homologue reached as low as 20-30% (Figure 11A). Since L17 is an essential r-protein, comparisons to a null mutant were not feasible, however, a substantial fitness reduction relative to the wild-type, was only observed when amino acid identity of the orthologue fell below 50%. Interestingly, higher fitness of certain L17 hybrid mutants relative to wild-type was also evident. A general trend that emerged was a reduction in relative fitness as phylogenetic distance increased, however the costs of replacement for the more homologous proteins were diminutive in comparison to the costs of removal.

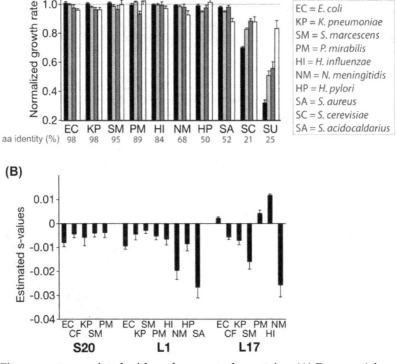

Fig. 11. **Fitness costs associated with replacement of r-proteins.** (A) Exponential growth rates of L1 replacement mutants in LB and minimal M9 medium with glucose, glycerol or acetate as carbon source. The growth rate of each mutant was normalized to that of a congenic wild-type strain in each respective medium (set to 1.0). (B) Selection coefficients associated with non-native S20, L1 and L17 r-proteins, estimated by means of pair-wise competitive growth with a congenic wild-type strain. Error bars in (A) and (B) represent standard error of the mean (SEM).

For those replacement mutants that displayed very small fitness costs as measured by exponential growth, pair-wise competition experiments were performed with the wild-type for a more sensitive estimate of the fitness costs (Lind et al., 2010). Using this method, we found that the majority of transfers were deleterious. In Figure 11B, the selection coefficient associated with each non-native protein is shown, which, in most cases, was less than or equal to -0.01, meaning that most of the imported genes conferred costs of 1% or less. Modelling was also performed to determine if such weak fitness costs represent an effective barrier for fixation of these proteins in nature via horizontal gene transfer. The results suggest that the fixation probability of such modestly counter-selected alleles is virtually nil in large natural populations of bacteria (Lind et al., 2010).

5.2 High functional conservation of r-proteins

Since the rate of bacterial growth is the product of the cellular ribosome concentration, times the rate of ribosome function (Bremer & Dennis, 1996), the fitness assays also provide an indirect measure of the translational capacity of the hybrid ribosomes. Given the generally larger fitness costs associated with complete loss of function in the null mutants, we were certain that the majority of non-native proteins improved fitness and thus appear to be at least semi-functional. Even though the replacement genes were often highly divergent in terms of base composition, codon usage and amino acid identity, overall, the fitness costs were relatively weak and significant reductions in fitness were only observed when the amino acid identity fell below 50% or so. So, although some of these proteins evolved in highly divergent organisms, they appear to function extraordinarily well in their non-native context. For example, L1 from *Helicobacter pylori* only shares 50% amino acid identity with the native *S. typhimurium* protein; the cost of replacement, however, is extremely small (approximately 1%). This clearly demonstrates the robustness of the ribosome and protein synthesis in general, and it also suggests that functional constraints are highly conserved between these proteins.

5.3 Increased dosage rescues suboptimal r-proteins

Those replacement mutants that harboured fitness costs of 10% or greater were subject to compensatory evolution to determine if and by what mechanisms fitness compensation could be achieved. After between 40 – 250 generations of growth, the S20 mutant (carrying the *H. influenzae* orthologue) and the L1 mutant (carrying the *S. cerevisiae* orthologue) showed increased fitness and were examined for an increase in the copy number of the non-native gene by Southern hybridization (Lind et al., 2010). Indeed, a two- to three-fold increase was detected in some of the lineages and corresponded to increased expression of the orthologous protein when Western blots were performed. To verify that overproduction of the non-native protein constitutes a general mechanism of fitness compensation, the orthologous proteins were expressed from an inducible high-copy number plasmid and fitness was measured. Exponential growth rates confirmed that for the majority of replacement mutants, increased dosage of the non-native protein conferred higher fitness (Figure 12). Thus, insufficient expression of the divergent protein (possibly due to unstable mRNA/protein), or a requirement for more of the protein to restore proper ribosome function and/or assembly, are two likely scenarios that could be at least partially accountable for the fitness costs of r-protein replacement. Such imbalances in the

Comprehending a Molecular Conundrum: Functional Studies of Ribosomal Protein Mutants from
Salmonella enterica Serovar Typhimurium

183

stoichiometry of interacting cellular components has previously been suggested to confer deleterious effects (Papp et al., 2003).

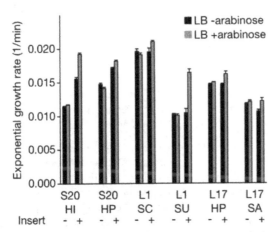

Fig. 12. **Growth compensation of r-protein replacement mutants via increased dosage of the non-native protein.** Exponential growth rates of mutants harbouring the arabinose inducible, high-copy number plasmid with (+) and without (-) the non-native gene. Abbreviations for divergent proteins from each respective species are the same as in Figure 11. Error bars represent SEM.

6. Concluding remarks and future perspectives

The ability to ascribe discrete functional roles to distinct components of the ribosome is essential to improve our understanding of how translation works and is a vital step towards furthering our knowledge of how this magnificent machine has evolved to the complex structure we see today in modern cells. The most fundamental aspect of the studies described herein, is directed towards resolving ribosome complexity to some degree in terms of the selective contributions made by certain r-proteins. The major tasks of the ribosome are RNA-mediated; however, the results of this study highlight the importance of the r-proteins for optimal function. Upon loss of either S20 or L1, we see that the cell is indeed viable; however, the substantial reductions in fitness reflect the requirement of the proteins for maintenance of proper ribosome function. In the case of both deletion mutants, the impairments in function appear to stem from perturbations in the higher order structure of the rRNA, suggesting, that both L1 and S20 are important in stabilizing the overall morphology of the subunits. Upon loss of S20, mRNA binding is severely impaired and association of the mutant 30S species with 50S subunits is drastically reduced. Both of these defects appear to occur as secondary effects caused by a primary disturbance in 16S rRNA structure; namely, distortion of the penultimate helix, h44. Similarly, in the absence of L1, an excess of free 50S subunits is formed in the cell, likely due to subunit misassembly in an L1-depleted environment. Moreover, the majority of compensatory mutations selected as a response to L1 removal were mapped to other r-proteins dispersed at various sites of the small and large subunit, suggesting that they may amend the structural perturbations to some degree. The variation in the topographical distribution of the compensatory mutations also highlights the high degree of co-operativity inherent to the ribosome. Thus, in both cases, it seems that loss of the protein indirectly perturbs protein synthesis via alterations in the overall morphology of the ribosome. Perhaps this is not altogether surprising, considering that both proteins are primary rRNA binders and are thus considered particularly important for initiating global folding of the major rRNA domains. One major strength and unique feature of the studies described herein, is our attempt to explain the observed fitness costs in terms of the costs to ribosome function detected by our *in vitro* system. Unfortunately, very few studies have attempted to bridge the crucial gap between both sets of data. This is a considerable challenge, given the inherent complexity of the ribosome and the enormous level of co-operativity within the structure itself and at the cellular level. It is, however, an essential step towards capturing a complete picture of the mechanism of translation.

In contrast to the substantial costs conferred by r-protein removal, in general, replacement of such proteins with orthologous counterparts from other species had only modest deleterious effects on fitness. This finding implies that the non-native proteins are capable of functioning in their new environment, despite having evolved in divergent species, clearly demonstrating the plasticity and robustness of the translation apparatus. This may suggest conservation in the overall shape of the proteins or at least those residues that interact with the target sites of the ribosome, akin to structural mimics of the native proteins. Overall the results emphasize the superiority of functional requirement over conservation of the primary sequence of amino acids, lending support to the suggestion that the major function of the r-proteins is to maintain the structural integrity of the ribosome.

Future work in this area should concentrate on studies designed to delineate the specific roles of other r-proteins during translation. With a large array of approximately fifty proteins, the task is challenging. However, recent advances in genetic engineering has allowed us to construct precise deletions of various r-proteins, and the replacement approach offers a new avenue for dissecting the roles of essential proteins. By examining the function of hybrid ribosomes *in vitro*, distinct steps of the translation cycle can be evaluated and the contributions made by essential r-proteins could be exposed. The findings of these studies also highlight the usefulness of compensatory evolution in terms of revealing otherwise concealed cooperative mechanisms for partially restoring ribosome function.

7. Acknowledgments

This work was supported by research grants to S.S. and D.A. from the Swedish Research Council and Knut and Alice Wallenberg Stiftelse; to S.S. from Carl Tryggers Stiftelse, Göran Gustafssons Stiftelse and Wenner-Gren Stiftelse.

8. References

Ameres, S. L., Shcherbakov, D., Nikonova, E., Piendl, W., Schroeder, R. and Semrad, K. (2007). RNA chaperone activity of L1 ribosomal proteins: phylogenetic conservation and splicing inhibition. *Nucleic Acids Res*, Vol. 35, No. 11, pp. 3752-3763

Baba, T., Ara, T., Hasegawa, M., Takai, Y., Okumura, Y., Baba, M., Datsenko, K. A., Tomita, M., Wanner, B. L. and Mori, H. (2006). Construction of Escherichia coli K-12 in-frame, single-gene knockout mutants: the Keio collection. *Mol Syst Biol*, Vol. 2, No. pp. 2006 0008

Ban, N., Nissen, P., Hansen, J., Moore, P. B. and Steitz, T. A. (2000). The complete atomic
    structure of the large ribosomal subunit at 2.4 A resolution. *Science*, Vol. 289, No.
    5481, pp. 905-920
Bjorkman, J., Samuelsson, P., Andersson, D. I. and Hughes, D. (1999). Novel ribosomal
    mutations affecting translational accuracy, antibiotic resistance and virulence of
    Salmonella typhimurium. *Mol Microbiol*, Vol. 31, No. 1, pp. 53-58
Bremer, H. and Dennis, P. P. (1996). Modulation of Chemical Composition and Other
    Parameters of the Cell by Growth Rate, In: *Escherichia coli and Salmonella: Cellular
    and Molecular Biology*, F. C. Neidhart, pp. 1553-1566, ASM Press, Washington D.C.
Brodersen, D. E., Clemons, W. M., Jr., Carter, A. P., Wimberly, B. T. and Ramakrishnan, V.
    (2002). Crystal structure of the 30 S ribosomal subunit from Thermus thermophilus:
    structure of the proteins and their interactions with 16 S RNA. *J Mol Biol*, Vol. 316,
    No. 3, pp. 725-768
Bubunenko, M., Baker, T. and Court, D. L. (2007). Essentiality of ribosomal and transcription
    antitermination proteins analyzed by systematic gene replacement in Escherichia
    coli. *J Bacteriol*, Vol. 189, No. 7, pp. 2844-2853
Bubunenko, M., Korepanov, A., Court, D. L., Jagannathan, I., Dickinson, D., Chaudhuri, B.
    R., Garber, M. B. and Culver, G. M. (2006). 30S ribosomal subunits can be
    assembled in vivo without primary binding ribosomal protein S15. *RNA*, Vol. 12,
    No. 7, pp. 1229-1239
Burgin, A. B., Parodos, K., Lane, D. J. and Pace, N. R. (1990). The excision of intervening
    sequences from Salmonella 23S ribosomal RNA. *Cell*, Vol. 60, No. 3, pp. 405-414
Cech, T. R. (2000). Structural biology. The ribosome is a ribozyme. *Science*, Vol. 289, No.
    5481, pp. 878-879
Cukras, A. R. and Green, R. (2005). Multiple effects of S13 in modulating the strength of
    intersubunit interactions in the ribosome during translation. *J Mol Biol*, Vol. 349,
    No. 1, pp. 47-59
Dabbs, E. R. (1977). A spectinomycin dependent mutant of Escherichia coli. *Mol Gen Genet*,
    Vol. 151, No. 3, pp. 261-267
Dabbs, E. R. (1978). Mutational alterations in 50 proteins of the Escherichia coli ribosome.
    Mol Gen Genet, Vol. 165, No. 1, pp. 73-78
Dabbs, E. R. (1979). Selection for Escherichia coli mutants with proteins missing from the
    ribosome. *J Bacteriol*, Vol. 140, No. 2, pp. 734-737
Dabbs, E. R. (1980). The ribosomal components responsible for kasugamycin dependence,
    and its suppression, in a mutant of Escherichia coli. *Mol Gen Genet*, Vol. 177, No. 2,
    pp. 271-276
Dabbs, E. R. (1991). Mutants lacking individual ribosomal proteins as a tool to investigate
    ribosomal properties. *Biochimie*, Vol. 73, No. 6, pp. 639-645
Dabbs, E. R. and Wittman, H. G. (1976). A strain of Escherichia coli which gives rise to
    mutations in a large number of ribosomal proteins. *Mol Gen Genet*, Vol. 149, No. 3,
    pp. 303-309
Dammel, C. S. and Noller, H. F. (1993). A cold-sensitive mutation in 16S rRNA provides
    evidence for helical switching in ribosome assembly. *Genes Dev*, Vol. 7, No. 4, pp.
    660-670
Datsenko, K. A. and Wanner, B. L. (2000). One-step inactivation of chromosomal genes in
    Escherichia coli K-12 using PCR products. *Proc Natl Acad Sci U S A*, Vol. 97, No. 12,
    pp. 6640-6645

Deutscher, M. P. (2009). Maturation and degradation of ribosomal RNA in bacteria. *Prog Mol Biol Transl Sci*, Vol. 85, No. pp. 369-391

Doudna, J. A. and Rath, V. L. (2002). Structure and function of the eukaryotic ribosome: the next frontier. *Cell*, Vol. 109, No. 2, pp. 153-156

Dutca, L. M. and Culver, G. M. (2008). Assembly of the 5' and 3' minor domains of 16S ribosomal RNA as monitored by tethered probing from ribosomal protein S20. *J Mol Biol*, Vol. 376, No. 1, pp. 92-108

Ehrenberg, M., Bilgin, N. and Kurland, C. G. (1989). Design and use of a fast and accurate in vitro translation system., In: *Ribosomes and Protein Synthesis: A Practical Approach*, G. Spedding, pp. 101-129, Oxford University Press, New York

Fei, J., Kosuri, P., MacDougall, D. D. and Gonzalez, R. L., Jr. (2008). Coupling of ribosomal L1 stalk and tRNA dynamics during translation elongation. *Mol Cell*, Vol. 30, No. 3, pp. 348-359

Giese, K. and Subramanian, A. R. (1991). Expression and functional assembly into bacterial ribosomes of a nuclear-encoded chloroplast ribosomal protein with a long NH2-terminal extension. *FEBS Lett*, Vol. 288, No. 1-2, pp. 72-76

Gotz, F., Dabbs, E. R. and Gualerzi, C. O. (1990). Escherichia coli 30S mutants lacking protein S20 are defective in translation initiation. *Biochim Biophys Acta*, Vol. 1050, No. 1-3, pp. 93-97

Gotz, F., Fleischer, C., Pon, C. L. and Gualerzi, C. O. (1989). Subunit association defects in Escherichia coli ribosome mutants lacking proteins S20 and L11. *Eur J Biochem*, Vol. 183, No. 1, pp. 19-24

Harms, J., Schluenzen, F., Zarivach, R., Bashan, A., Gat, S., Agmon, I., Bartels, H., Franceschi, F. and Yonath, A. (2001). High resolution structure of the large ribosomal subunit from a mesophilic eubacterium. *Cell*, Vol. 107, No. 5, pp. 679-688

Johansson, M., Bouakaz, E., Lovmar, M. and Ehrenberg, M. (2008). The kinetics of ribosomal peptidyl transfer revisited. *Mol Cell*, Vol. 30, No. 5, pp. 589-598

Kaczanowska, M. and Ryden-Aulin, M. (2007). Ribosome biogenesis and the translation process in Escherichia coli. *Microbiol Mol Biol Rev*, Vol. 71, No. 3, pp. 477-494

Korepanov, A. P., Gongadze, G. M., Garber, M. B., Court, D. L. and Bubunenko, M. G. (2007). Importance of the 5 S rRNA-binding ribosomal proteins for cell viability and translation in Escherichia coli. *J Mol Biol*, Vol. 366, No. 4, pp. 1199-1208

Kramer, E. B. and Farabaugh, P. J. (2007). The frequency of translational misreading errors in E. coli is largely determined by tRNA competition. *RNA*, Vol. 13, No. 1, pp. 87-96

Lecompte, O., Ripp, R., Thierry, J. C., Moras, D. and Poch, O. (2002). Comparative analysis of ribosomal proteins in complete genomes: an example of reductive evolution at the domain scale. *Nucleic Acids Res*, Vol. 30, No. 24, pp. 5382-5390

Lind, P. A., Tobin, C., Berg, O. G., Kurland, C. G. and Andersson, D. I. (2010). Compensatory gene amplification restores fitness after inter-species gene replacements. *Mol Microbiol*, Vol. 75, No. 5, pp. 1078-1089

Liu, X. Q., Gillham, N. W. and Boynton, J. E. (1989). Chloroplast ribosomal protein gene rps12 of Chlamydomonas reinhardtii. Wild-type sequence, mutation to streptomycin resistance and dependence, and function in Escherichia coli. *J Biol Chem*, Vol. 264, No. 27, pp. 16100-16108

Mandava, C. S., Peisker, K., Ederth, J., Kumar, R., Ge, X., Szaflarski, W., and Sanyal, S (2011). Bacterial ribosome requires multiple L12 dimers for efficient initiation and elongation of protein synthesis involving IF2 and EF-G. *Nucleic Acid Research*, doi: 10.1093/nar/gkr1031.

Mattatall, N. R. and Sanderson, K. E. (1996). Salmonella typhimurium LT2 possesses three distinct 23S rRNA intervening sequences. *J Bacteriol*, Vol. 178, No. 8, pp. 2272-2278

Mattatall, N. R. and Sanderson, K. E. (1996). Salmonella typhimurium LT2 possesses three distinct 23S rRNA intervening sequences. *J Bacteriol*, Vol. 178, No. 8, pp. 2272-2278

Mattatall, N. R. and Sanderson, K. E. (1998). RNase III deficient Salmonella typhimurium LT2 contains intervening sequences (IVSs) in its 23S rRNA. *FEMS Microbiol Lett*, Vol. 159, No. 2, pp. 179-185

McClelland, M., Sanderson, K. E., Spieth, J., Clifton, S. W., Latreille, P., Courtney, L., Porwollik, S., Ali, J., Dante, M., Du, F., Hou, S., Layman, D., Leonard, S., Nguyen, C., Scott, K., Holmes, A., Grewal, N., Mulvaney, E., Ryan, E., Sun, H., Florea, L., Miller, W., Stoneking, T., Nhan, M., Waterston, R. and Wilson, R. K. (2001). Complete genome sequence of Salmonella enterica serovar Typhimurium LT2. *Nature*, Vol. 413, No. 6858, pp. 852-856

Miller, J. H. (1992). *A Short Course in Bacterial Genetics*, Cold Spring Harbour Laboratory Press, Cold Spring Harour, New York

Munro, J. B., Altman, R. B., Tung, C. S., Cate, J. H., Sanbonmatsu, K. Y. and Blanchard, S. C. (2010). Spontaneous formation of the unlocked state of the ribosome is a multistep process. *Proc Natl Acad Sci U S A*, Vol. 107, No. 2, pp. 709-714

Nierhaus, K. H. (1991). The assembly of prokaryotic ribosomes. *Biochimie*, Vol. 73, No. 6, pp. 739-755

Nierhaus, K. H. (2006). Bacterial Ribosomes: Assembly, In: *Encyclopedia of Life Sciences*, pp. Pages, John Wiley & Sons Ltd., Retrieved from http://www.els.net/

Nissen, P., Hansen, J., Ban, N., Moore, P. B. and Steitz, T. A. (2000). The structural basis of ribosome activity in peptide bond synthesis. *Science*, Vol. 289, No. 5481, pp. 920-930

O'Connor, M. and Gregory, S. T. (2011). Inactivation of the RluD pseudouridine synthase has minimal effects on growth and ribosome function in wild-type Escherichia coli and Salmonella enterica. *J Bacteriol*, Vol. 193, No. 1, pp. 154-162

Ogle, J. M., Carter, A. P. and Ramakrishnan, V. (2003). Insights into the decoding mechanism from recent ribosome structures. *Trends Biochem Sci*, Vol. 28, No. 5, pp. 259-266

Papp, B., Pal, C. and Hurst, L. D. (2003). Dosage sensitivity and the evolution of gene families in yeast. *Nature*, Vol. 424, No. 6945, pp. 194-197

Pettersson, I. and Kurland, C. G. (1980). Ribosomal protein L7/L12 is required for optimal translation. *Proc Natl Acad Sci U S A*, Vol. 77, No. 7, pp. 4007-4010

Ryden-Aulin, M., Shaoping, Z., Kylsten, P. and Isaksson, L. A. (1993). Ribosome activity and modification of 16S RNA are influenced by deletion of ribosomal protein S20. *Mol Microbiol*, Vol. 7, No. 6, pp. 983-992

Sander, G. (1983). Ribosomal protein L1 from Escherichia coli. Its role in the binding of tRNA to the ribosome and in elongation factor g-dependent gtp hydrolysis. *J Biol Chem*, Vol. 258, No. 16, pp. 10098-10103

Schaub, R. E. and Hayes, C. S. (2011). Deletion of the RluD pseudouridine synthase promotes SsrA peptide tagging of ribosomal protein S7. *Mol Microbiol*, Vol. 79, No. 2, pp. 331-341

Schluenzen, F., Tocilj, A., Zarivach, R., Harms, J., Gluehmann, M., Janell, D., Bashan, A., Bartels, H., Agmon, I., Franceschi, F. and Yonath, A. (2000). Structure of functionally activated small ribosomal subunit at 3.3 angstroms resolution. *Cell*, Vol. 102, No. 5, pp. 615-623

Selmer, M., Dunham, C. M., Murphy, F. V. t., Weixlbaumer, A., Petry, S., Kelley, A. C., Weir, J. R. and Ramakrishnan, V. (2006). Structure of the 70S ribosome complexed with mRNA and tRNA. *Science*, Vol. 313, No. 5795, pp. 1935-1942

Sharan, S. K., Thomason, L. C., Kuznetsov, S. G. and Court, D. L. (2009). Recombineering: a homologous recombination-based method of genetic engineering. *Nat Protoc*, Vol. 4, No. 2, pp. 206-223

Subramanian, A. R. and Dabbs, E. R. (1980). Functional studies on ribosomes lacking protein L1 from mutant Escherichia coli. *Eur J Biochem*, Vol. 112, No. 2, pp. 425-430

Tobin, C., Mandava, C. S., Ehrenberg, M., Andersson, D. I. and Sanyal, S. (2010). Ribosomes lacking protein S20 are defective in mRNA binding and subunit association. *J Mol Biol*, Vol. 397, No. 3, pp. 767-776

Trabuco, L. G., Schreiner, E., Eargle, J., Cornish, P., Ha, T., Luthey-Schulten, Z. and Schulten, K. (2010). The role of L1 stalk-tRNA interaction in the ribosome elongation cycle. *J Mol Biol*, Vol. 402, No. 4, pp. 741-760

Tubulekas, I., Buckingham, R. H. and Hughes, D. (1991). Mutant ribosomes can generate dominant kirromycin resistance. *J Bacteriol*, Vol. 173, No. 12, pp. 3635-3643

Uchiumi, T., Honma, S., Nomura, T., Dabbs, E. R. and Hachimori, A. (2002). Translation elongation by a hybrid ribosome in which proteins at the GTPase center of the Escherichia coli ribosome are replaced with rat counterparts. *J Biol Chem*, Vol. 277, No. 6, pp. 3857-3862

Valle, M., Zavialov, A., Sengupta, J., Rawat, U., Ehrenberg, M. and Frank, J. (2003). Locking and unlocking of ribosomal motions. *Cell*, Vol. 114, No. 1, pp. 123-134

Weglohner, W., Junemann, R., von Knoblauch, K. and Subramanian, A. R. (1997). Different consequences of incorporating chloroplast ribosomal proteins L12 and S18 into the bacterial ribosomes of Escherichia coli. *Eur J Biochem*, Vol. 249, No. 2, pp. 383-392

Wild, D. G. (1988). Reversion from erythromycin dependence in Escherichia coli: strains altered in ribosomal sub-unit association and ribosome assembly. *J Gen Microbiol*, Vol. 134, No. 5, pp. 1251-1263

Williamson, J. R. (2009). The ribosome at atomic resolution. *Cell*, Vol. 139, No. 6, pp. 1041-1043

Wilson, D. N. and Nierhaus, K. H. (2005). Ribosomal proteins in the spotlight. *Crit Rev Biochem Mol Biol*, Vol. 40, No. 5, pp. 243-267

Wilson, D. N. and Nierhaus, K. H. (2006). The E-site story: the importance of maintaining two tRNAs on the ribosome during protein synthesis. *Cell Mol Life Sci*, Vol. 63, No. 23, pp. 2725-2737

Wimberly, B. T., Brodersen, D. E., Clemons, W. M., Jr., Morgan-Warren, R. J., Carter, A. P., Vonrhein, C., Hartsch, T. and Ramakrishnan, V. (2000). Structure of the 30S ribosomal subunit. *Nature*, Vol. 407, No. 6802, pp. 327-339

Wower, I. K., Wower, J. and Zimmermann, R. A. (1998). Ribosomal protein L27 participates in both 50 S subunit assembly and the peptidyl transferase reaction. *J Biol Chem*, Vol. 273, No. 31, pp. 19847-19852

Yusupov, M. M., Yusupova, G. Z., Baucom, A., Lieberman, K., Earnest, T. N., Cate, J. H. and Noller, H. F. (2001). Crystal structure of the ribosome at 5.5 A resolution. *Science*, Vol. 292, No. 5518, pp. 883-896

Permissions

The contributors of this book come from diverse backgrounds, making this book a truly international effort. This book will bring forth new frontiers with its revolutionizing research information and detailed analysis of the nascent developments around the world.

We would like to thank Bassam A. Annous and Joshua B. Gurtler, for lending their expertise to make the book truly unique. They have played a crucial role in the development of this book. Without their invaluable contribution this book wouldn't have been possible. They have made vital efforts to compile up to date information on the varied aspects of this subject to make this book a valuable addition to the collection of many professionals and students.

This book was conceptualized with the vision of imparting up-to-date information and advanced data in this field. To ensure the same, a matchless editorial board was set up. Every individual on the board went through rigorous rounds of assessment to prove their worth. After which they invested a large part of their time researching and compiling the most relevant data for our readers. Conferences and sessions were held from time to time between the editorial board and the contributing authors to present the data in the most comprehensible form. The editorial team has worked tirelessly to provide valuable and valid information to help people across the globe.

Every chapter published in this book has been scrutinized by our experts. Their significance has been extensively debated. The topics covered herein carry significant findings which will fuel the growth of the discipline. They may even be implemented as practical applications or may be referred to as a beginning point for another development. Chapters in this book were first published by InTech; hereby published with permission under the Creative Commons Attribution License or equivalent.

The editorial board has been involved in producing this book since its inception. They have spent rigorous hours researching and exploring the diverse topics which have resulted in the successful publishing of this book. They have passed on their knowledge of decades through this book. To expedite this challenging task, the publisher supported the team at every step. A small team of assistant editors was also appointed to further simplify the editing procedure and attain best results for the readers.

Our editorial team has been hand-picked from every corner of the world. Their multi-ethnicity adds dynamic inputs to the discussions which result in innovative outcomes. These outcomes are then further discussed with the researchers and contributors who give their valuable feedback and opinion regarding the same. The feedback is. then collaborated with the researches and they are edited in a comprehensive manner to aid the understanding of the subject.

Apart from the editorial board, the designing team has also invested a significant amount of their time in understanding the subject and creating the most relevant covers. They scrutinized every image to scout for the most suitable representation of the subject and create an appropriate cover for the book.

The publishing team has been involved in this book since its early stages. They were actively engaged in every process, be it collecting the data, connecting with the contributors or procuring relevant information. The team has been an ardent support to the editorial, designing and production team. Their endless efforts to recruit the best for this project, has resulted in the accomplishment of this book. They are a veteran in the field of academics and their pool of knowledge is as vast as their experience in printing. Their expertise and guidance has proved useful at every step. Their uncompromising quality standards have made this book an exceptional effort. Their encouragement from time to time has been an inspiration for everyone.

The publisher and the editorial board hope that this book will prove to be a valuable piece of knowledge for researchers, students, practitioners and scholars across the globe.

List of Contributors

Madalena Vieira-Pinto
Departamento das Ciências Veterinárias, CECAV Laboratório de TPA & Inspecção Sanitária, Universidade de Trás-os-Montes e Alto Douro, Portugal

Patrícia Themudo
Laboratório Nacional de Investigação Veterinária, Lisboa, Portugal

Lucas Dominguez, José Francisco Fernandez-Garayzabal and Ana Isabel Vela
Centro de Vigilancia Sanitaria Veterinaria (VISAVET), Universidad Complutense, Madrid, España

José Francisco Fernandez-Garayzabal and Ana Isabel Vela
Departamento de Sanidad Animal, Facultad de Veterinaria, Universidad Complutense, Madrid, España

Fernando Bernardo, Cristina Lobo Vilela and Manuela Oliveira
Centro Interdisciplinar de Investigação em Sanidade Animal, Faculdade de Medicina Veterinária da Universidade, Técnica de Lisboa, Lisboa, Portugal

Rosselin Manon, Abed Nadia, Namdari Fatémeh, Virlogeux-Payant Isabelle, Velge Philippe and Wiedemann Agnès
INRA Centre de Tours, UR1282 Infectiologie Animale et Santé Publique, Nouzilly, IFR 136, Agents Transmissibles et Infectiologie, Nouzilly, France

Liqing Zhou, Thomas Darton, Claire Waddington and Andrew J. Pollard
University of Oxford, United Kingdom

Robert S. Tebbs, Lily Y. Wong, Pius Brzoska and Olga V. Petrauskene
Life Technologies, Foster City, CA, USA

R. Margaret Wallen and Michael H. Perlin
University of Louisville, USA

Chantal G. Forest and France Daigle
Department of Microbiology & Immunology/University of Montreal, Canada

Yusuke V. Morimoto and Tohru Minamino
Osaka University, Japan

Christina Tobin Kåhrström and Dan I. Andersson
Department of Medical Biochemistry and Microbiology, Sweden

Suparna Sanyal
Department of Cell and Molecular Biology, Uppsala University, Sweden